CANOE COUNTRY

Plants and Trees of the
North Woods and Boundary Waters

Mark Stensaas

Illustrations by Jeff Sonstegard

University of
Minnesota Press

Minneapolis • London

Originally published by Pfeifer-Hamilton Publishers, 1996
First University of Minnesota Press edition, 2004

Published by the University of Minnesota Press
111 Third Avenue South, Suite 290
Minneapolis, MN 55401-2520
http://www.upress.umn.edu

ISBN 0-8166-4503-5 (PB)

A Cataloging-in-Publication record for this book is available from the Library of Congress.

Printed in the United States of America on acid-free paper

The University of Minnesota is an equal-opportunity educator and employer.

12 11 10 09 08 07 06 05 04   10 9 8 7 6 5 4 3 2 1

Bill –
Thanks for sharing
your passion for the
BWCAW and canoeing with
visitors to this special
part of the world.
And welcome to "professional"
guiding!
Cheers,
Jim

CANOE COUNTRY

Flora

Plants and Trees
of the North Woods
and Boundary Waters

"To God be the glory,
great things He hath done"

CONTENTS

Trees

Shrubs

Wildflowers

Wildflowers

Ferns and Other Nonflowering Plants

Ferns

Clubmosses

Horsetails

Mosses

Lichens

Fungi

Appendix

Acknowledgments

Like the floral community of the North Woods ecosystem, a book is also a complex web of interconnected parts that when working together result in a beautiful whole. This book has been one such project. Every person who made a contribution to *Canoe Country Flora* was crucial to the finished product.

A big thanks to Ann Robertson of North Angle Writing for letting me share her tiny office and computer. I wrote my first book in pencil, and what a difference a computer made. I can't even remember life before spellcheck now. For giving me the time I needed to write this book, I must thank Cindy Emerson, my boss at Duluth Pack. It would not have been possible without the perk of the 90s—flexible scheduling.

Much of the scientific data used in the book was gleaned from current botanical research. Without years of fieldwork and lab time by dedicated botanists, ecologists, dendrologists, mycologists, and others, this book would have been impossible. Equally important is the early work of ethnologists such as Frances Densmore, who wrote extensively during the early part of this century on cultural and medicinal uses of plants by the Ojibwa of Minnesota.

Valuable medical information came from three friends: Dr. Nancy Olsen, midwife Leslie Goodell, and midwife Cindy Anderson. Janet Pettersen found for me some fantastic lichen resources. While in Sweden, Jan Christoffersson and Anne-Soifi Modig helped me translate Swedish plant names. For phenology data, I'm indebted to John Latimer of Grand Rapids, Ken and Molly Hoffman of Grand Marais, Chel Anderson of Tofte, and Tammy Rick of Litchfield. And thanks to Tobin Schmuck for a warm place to work when all I had was a cold cabin.

My scientific reviewers thoroughly perused the text to make sure all the facts were in order and accurate. A huge hoorah for Chel Anderson, The Nature Conservancy botanist; Bunter Knowles, naturalist/nurse; and Cal Harth, ecologist/naturalist/ornithologist/botanist for the Natural Resources Research Institute for going over the entire manuscript with a fine-toothed comb. Dave Roon, amateur mycologist, checked over my fungi chapters. Good friend Timo Rova was my lay reviewer, checking the chapters for "leaps of logic."

We all owe Pfeifer-Hamilton Publishers a big North Woods cheer for putting out great regional books by talented local authors. Founder Don Tubesing gave me my first shot at writing a book, and now he's letting me do it again. Thanks, Don. Without Susan Gustafson's patience and beyond-the-call-of-duty effort, *Canoe Country Flora* could not have been completed . . . at least on time anyway. Joy Morgan Dey is a talented graphic artist who designed the fantastic layout and eye-catching cover for this book. Kathy DeArmond-Lundblad put up with my last minute changes and tweaked the text into fine shape.

Editor Carol Kennedy straightened up my syntax, cleaned out the confusing clutter, purified my paragraphs, tightened up the text, and in general choreographed my chaos into an easily read work.

And finally, I am extremely grateful for the work of my talented artist Jeff Sonstegard, who died suddenly this fall. I'd seen his work in other Pfeifer-Hamilton books and was impressed. But I gave him Mission Impossible. I needed one hundred very detailed drawings in three months. He smiled and did it in a month and a half despite a full-time job. His work is superb and brings life to the text. His art will be a living testament to the man. We all miss Jeff.

"Sparky"
Mark Stensaas
October 1995

Introduction—
The North Woods

I grew up in suburban Minneapolis, where "wild" was the brushy right-of-way along the railroad. Wild was Ring-necked Pheasants exploding from underfoot. Wild was wild grapes and the tall exotic-looking mullein plants. (They made great swords!) My first trip to the North Woods was in high school. I'd read several books by the great philosopher-naturalist-author Sigurd Olson, and boy, was I pumped. It was a church trip to Minnesota's Boundary Waters Canoe Area Wilderness, and I'll never forget it. I'll never forget it because I was so disappointed. Water, rocks, and trees, that's all it was. No wait . . . water, rocks, trees, and mosquitoes. "This North Woods stuff is boring," I remember thinking. But later that summer I discovered flowers and began keying out the flora of suburbia. I bought a *Peterson Field Guide to Wildflowers—Northeast Region.*

When I returned to the North Woods, it was with new eyes. The North is a subtle place . . . superficially lifeless and sterile . . . but when you delve into the lives of the plants that grow there, you find a wonderfully diverse and fantastically created green universe.

It's not an easy place to live. Winter is long, the snow deep. Summer can be blistering hot. Violent storms rumble through. Winds desiccate, uproot, and beat down.

But what defines the "North Woods"? Henry David Thoreau described it as a place where the woods are all "sprucey and moosey." Moose, though, wander through the forests and farmland of North Dakota also. And what about the Engelmann Spruce (*Picea engelmannii*) which grows at high elevations in Arizona? Is that North Woods? What images come to mind when you think of the North Woods? Log cabins? Lumberjacks? Pines? Lakes? Visions of the "sky blue waters" of Minnesota may pop into your head. Maine, Michigan, Manitoba, Algonquin, Adirondack, birchbark canoes, summer camp, Hiawatha, mosquitoes, blueberries, Bald Eagles, Black Bears, blackflies, Black Spruce? How about Duluth Packs, Old Towne wood-and-canvas canoes, ash pack baskets, snowshoes, skis, fishing, L.L. Bean, Paul Bunyan and Babe the Blue Ox? Maybe Finns, Norwegians, the Ojibwa, Penobscots, Iroquois, Northern Pike, Walleye, Lake Trout, logging, shipping, mining, iron ore, gold, silver, copper, granite. . . . wait!

All these things are the North Woods to different people. But for the purposes of this book we must biologically define this region by geologic history, soils, climate, and dominant flora.

From a million years ago to roughly twelve thousand years ago, mile-high moving monsters of ice—glaciers—ground it down and scraped it clean like a giant bulldozer. This latest episode of purging by ice happened so geologically recently that the earth is still rebounding from the weight that's now gone. Lichens came first to the clean white granite and other bedrocks. Returning birds and mammals carried seeds in their feces, feathers, and fur. Other seeds and spores literally blew into town, riding the wind to a new life in a strange place. Things grew. Things died. Soil was built. Larger plants and trees could now survive here.

Glacial meltwater trapped in scoured rock depressions grew over with sphagnum mosses and slowly filled downward forming bogs, peatlands, and fens. The labyrinth of lakes and rivers that one finds in the North is a direct result of the ice and bedrock. It is said that on the Canadian Shield you are never more than a mile away from navigable water. The land is literally water, rocks, trees, and sky.

But creating dirt is a slow process. During the brief span since the glaciers receded, only a thin skin of soil has built up. Today, an average of a mere six inches of soil covers the bedrock of northern Minnesota's border county. What grows in and on that soil is determined by rainfall, temperature, fire frequency, and soil nutrients. Plants, in order to survive, must endure summer temperatures of 100° Fahrenheit, winter temperatures of -60° Fahrenheit, drying winter winds, and low nutrient soils. The majority of plants are evergreen which saves growth time and energy during the brief summers.

The area in North America that fits our definition of North Woods forms a giant ring about Hudson Bay taking in the southern half of Labrador, Quebec, Ontario, Manitoba, and northern Alberta and dipping south of the border to take in northeast Minnesota, northern Wisconsin, the Upper Peninsula of Michigan, upstate New York, northern New England, and higher elevations in the Appalachians. Also known as the boreal and subboreal forest this will be our North Woods world.

This friendly field guide

Canoe Country Flora is a cultural and natural history guide to the most common trees, shrubs, wildflowers, ferns, mosses, club-mosses, horsetails, lichens, and fungi of the North Woods. Though written specifically for the wilds of northern Minnesota, this guide is just as easily at home in the entire area just defined as the North Woods.

The ninety-six species of native flora and fungi included in this book are those you are most likely to encounter on a canoe trip, backpacking excursion, day hike, on your way to the outhouse, or just around your cabin. In fact, I'm willing to wager that these species combine to make up over 90 percent of the vegetative mass in the North Woods. Of course there are thousands of plant species out there, so you will encounter a number that are not included in this book. For those, you'll have to do your own research.

The format

Species are grouped together in the chapters by kind. To facilitate easy reference, the information for each species is presented in a consistent format.

Quick identifiers—Accurate line drawings and the brief descriptions that accompany them will help you rapidly identify the plants you discover.

Natural history—Information about the plant's physical characteristics, preferred habitat, and life cycle, in addition to historical medicinal remedies, Native American uses, food uses, and other interesting scientific research will give you a deeper knowledge of the species. Caution: Obtain the advice of experts before using wild plants as food or medicine. Some of these plants contain powerful compounds that could be dangerous.

Tidbits—Fascinating little-known facts are highlighted in the sidebars.

Activities—The "Sparky says" activities will help you learn firsthand the variety and wonderment of our northern flora.

Phenology chart—This handy graph on page 200 shows you when the wildflowers are in bloom, enabling your identifications to be that much more sure.

Checklists—Species checklists are included at the end of each section. An additional checklist on pages 198 and 199 groups species by habitat.

Style—In keeping with the informal style of this guide, common names are used in the headings and in the text. Those common names referring to a single species are capitalized, while generic common names are left in lower case. Since common names vary regionally, the scientific binomial Latin name for each species is also included. Species without a chapter in the book are identified by common name and Latin name when first introduced, after which they are just referred to by common name. Accepted common and scientific names were taken from *Manual of Vascular Plants* (Gleason and Cronquist), *Grays Manual of Botany* (Fernald edition), *A Field Guide to Wildflowers* (Peterson), and *Vascular Plants of Minnesota* (Ownbey and Morley).

A final note

So it's time to get down on your knees and discover the "Canadian carpet." Take notes, take photos, make sketches, smell, nibble, feel. Do the "Sparky says" activities. Use the checklists. But most of all use the book. Take it along. Cram it in your trusty Duluth Pack, roll it up and stuff it in your fanny pack, strap it to your mountain bike. But just get out there and into the North Woods!

Trees

Balsam Fir

Abies balsamea

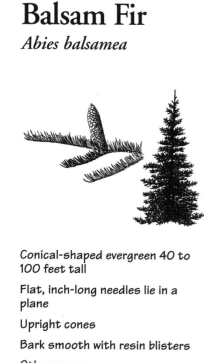

Conical-shaped evergreen 40 to 100 feet tall

Flat, inch-long needles lie in a plane

Upright cones

Bark smooth with resin blisters

Other names

    ginggop (Ojibwa: "ging-GOP")

    Canada balsam

    blister fir

    balm of Gilead fir

    silver pine

    sapin (Quebec)

Balsam Fir often gets a bum rap as being the "weed of the North." Because it is little used as lumber and susceptible to disease, many scorn the fragrant and short-lived *Abies balsamea*. It is lucky to hit eighty years. Abundant in the North Woods, it is considered part of the climax forest, the one that if left unaltered would regenerate itself to the end of time. Climax Black Spruce–Balsam Fir–Jack Pine forests are mysterious places where the "Moose and the Pine Marten play."

Balsam Fir is a conical evergreen with branches reaching near to or all the way to the ground. Its needles are flat—shiny green above and hoary below with two white stripes. To help identify this tree, look for these "racing stripes" and think of a sports car . . . like a "Fir-arri." (Hey, it works for me!) The needles (½ to 1 inch) tend to lie in a flat horizontal plane, but on closer examination you'll notice that they are attached around the entire stem. Plucking a needle off the stem leaves a smooth surface. Spruce needles, on the other hand, are attached to a ridged stem by a tiny stalk. Crushing a Balsam needle releases a fragrant rush of the essence of the North Woods. The bark of mature specimens is relatively smooth but pimpled with resin blisters. Cones, as on all firs, grow erect on the upper branches. As the cone scales fall, so do the seeds, leaving only the short spikelike cone stalk. Spruce cones hang from the branches.

Canada's Yukon Territories (the southern corner) and Newfoundland are the northern limits for Balsam Fir. It is common throughout the Great Lakes region and New England, dipping south to West Virginia along the Appalachians. Young firs growing up through the understory beneath mature Paper Birch are a common sight in the North. Thriving in the shade, the Balsams will eventually take over the canopy and become part of the climax forest—until, that is, a Spruce Budworm outbreak or forest fire kills them. It also frequently associates with White Spruce, Quaking Aspen, and Jack Pine. Old-growth trees may have a diameter at breast height (dbh) of 18 inches and tower to 100 feet. Eighty years old qualifies a Balsam for senior citizen status.

The Christmas connection

Balsam Fir is shaped just like the Christmas trees you drew as a kid. In fact, various sources report that 13 to 30 percent of all trees cut for the winter Yuletide are Balsams. "Far better that the little tree should arrive, like a shining child at your door, breathing of all out of doors and cupping healthy North Woods cold between its boughs, to bring delight to human children," expounds Donald Culross Peattie on their use as Christmas trees. To crush a Balsam needle is to smell all the holiday memories of childhood.

Origins of its link to Christmas can be traced to Germanic tribes, who practically worshipped the lowly tree. The great Christian reformer Martin Luther was allegedly the first person to put candles on a fir, and hence he "invented" the Christmas tree. The candles may have been imitations of the upright bare cone stalks tipped with snow. "Oh Tannenbaum," that holiday favorite, refers to firs used for the same purpose.

Since Balsams can have sparse branches, commercial growers encourage fullness by scoring the bark to stimulate bud development. Balsams, besides being very fragrant, also hold their needles longer than spruce.

Homegrown pesticides

Only every two to four years does Balsam Fir produce an abundant cone crop. It is therefore very protective of its seed. In 1962, a forest researcher published a report speculating that toxins were responsible for the low rate of Balsam seeds cached by White-footed Mice, Red-backed Voles, and Meadow Voles. The resin in the seeds also discourages predation by Red Squirrels, White-winged Crossbills, and Red Crossbills.

Insects that feed on needles also run into a secret weapon of the Balsam. Evidently, chemicals manufactured by the needles mimic growth hormones of several insect herbivores. Ingesting them interferes with metamorphosis, and the pesky critters fail to fully develop. Unfortunately for the Balsam, this chemical

Blister buster

"Canada balsam" is a clear liquid oleoresin found in the bark blisters on Balsam Fir trunks. At one time Maine loggers and local Indians used it as an emergency salve for cuts. Moving from lumberjack hands to delicate optical equipment, this same resin has also been used to cement glass lenses in microscopes and to secure specimens on glass microscope slides. It is pure and transparent and refracts light to the same degree as glass.

Spruce Budworm actually prefer to feed on Balsam Fir needles.

warfare is ineffective against its worst nightmare, the Spruce Budworm.

The plague

The Spruce Budworm (*Choristoneura fumiferana*), despite its name, unleashes its destructive appetite on large stands of Balsam Fir. Actually a caterpillar, the Spruce Budworm voraciously consumes needles of any and all firs in its path. To reach a new feeding site, the inch-long critters rappel off branch tips on a silken rope. At the point when you're picking them out of your hair every time you go outside, they get really annoying to humans like me. In early July they spin a loose cocoon in bunches of needles. Hundreds may infest a single tree. By this time the Balsams are a sorry lot, partially or entirely defoliated and covered with messy cocoons. The adult, a small gray moth, emerges in late July and lays the seeds of next year's plague on the needles of Balsam, Black Spruce, and White Spruce. In about a week, the tiny caterpillars emerge and promptly make for the nearest bark crevice, where they hibernate through the winter.

Infestations are usually massive. Minnesota's 1984 outbreak defoliated 150,000 acres and affected millions more. Outbreaks usually last three to five years and kill millions of trees, opening the door for massive forest fires. Mortality of Balsams may be as high as 100 percent in mature spruce-fir forests. A 1977 to 1982 study on a 186,000-acre Spruce Budworm-damaged tract reported that a half million cords of Balsam Fir and eight thousand cords of White Spruce were lost. But calculated out, this is only one-half cord per acre per year—in other words, not much. Pesticide control is a ridiculous option since it would mean spending millions to protect a resource that is not even used. Besides, Spruce Budworm is as North American as apple pie and Chevrolet. It is not some alien pest with no natural enemies and no place in the dynamics of the boreal ecosystem. Populations of Bay-breasted, Cape May, and Tennessee Warblers all increase dramatically during such outbreaks.

Ojibwa medicine

The Balsam Fir was a veritable medicine chest to the Lake Superior Ojibwa. Twigs were steeped in water to make a laxative. Bits of root were placed in the mouth to cure sores. Needles were tossed onto the sweat lodge's rocks to release a fragrant steam that cleared nasal passages. Gonorrhea was allayed by a warm liquid made from the sap. Inner bark had several uses. A tea of it eased chest pain, and the inhaled steam mixed with tobacco sped childbirth.

Sparky says: Go ahead! Pop a few of those tempting bark blisters. Pick a big fat one, and squeeze one side with your thumb. Be careful, though, as the resin can shoot ten or twenty feet, and it doesn't come out of clothes.

Now grab a twig or needle and dip one end in the gooey resin. Gently place it in the water and watch it amazingly putt along. Notice the "oil slick" created on the water's surface by the resin's constituents.

Black Spruce
Picea mariana

Scraggly and very pointed spruce of the sphagnum bogs

Short needles are diamond-shaped in cross section

Branches droop to the ground

1/2- to 1 1/2-inch cones cluster in tree top

Other names

gawandug (Ojibwa: "ga-WA-n-DUG")

bog spruce

swamp spruce

double spruce

epinette noir (Quebec)

Black Spruce is a tree of the wild and lonely lands. One author observantly noticed that in Minnesota and Canada the Black Spruce is a "dominant tree over a vast area which you can locate at once on the map by the fact that there are almost no towns, no roads, no railroads." Its penchant for the boggy, soggy sphagnum moss mats limits human encroachment. Today, though, the booming paper pulp industry generates a high demand for spruce. Billions of board feet per year are chewed up to make glossy paper advertising supplements and magazines.

Normally a short, scraggly tree, Black Spruce grows to its vertical zenith in the low hills of Manitoba and Saskatchewan, where it may reach 100 feet. By contrast, a century-old specimen growing in an acidic bog may be stunted to 10 feet. Drooping branches create a spiky-looking tree. Bark is thin and composed of numerous scalelike plates. Most trees are tipped by a dense head of short branches and cones, the weight of which, especially under a snow load, can snap the top off. The cones are small (1/2 to 1 1/2 inch) and soft, and shed whole, unlike the break-apart cones of Balsam Fir. They are semiserotinous (partially closed by resin), which provide them with some protection from forest fires that sweep the North.

Needles are short and stiff with a sharp point. Attached to the ridged stem by a short woody stalk, they sweep forward on the stem. Unlike the flat Balsam Fir needles, spruce needles are diamond-shaped in cross section, making it possible to roll them between your fingers.

White Spruce (*Picea glauca*) is an upland tree that usually grows to much larger dimensions (2 feet in diameter and 100 feet tall) than the bog-loving Black. The White's 1-inch-long needles are twice as long as the 1/2-inch needles of Black Spruce.

Step aside, Wrigley's

Spruce gum was a hot item in the turn-of-the-century lumber camps. Collectors scoured the woods with long-handled chisels, scraping the resinous exudations from the bark, and turned around and sold the stuff as chewing gum to the lumberjacks.

In 1862, Canadian trader Bernard Ross wrote of its use by local Ojibwa: "The gum is chewed by the female aborigines to the whiteness of whose teeth the habit contributes in no small degree." The spruce gum, molded into sticks and sold in stores, became all the rage in nineteenth-century America. This home-grown chew was replaced by tropical chicle in the twentieth century. "Chiclets" drew its name from this substance.

Wilderness thread

"Wattape" is the Cree name for spruce roots. Many tribes gathered the fine roots for use in sewing birchbark canoes and large baskets. Stored in bundles, the wattape was soaked and split before being used.

Sparky says: For that bright white smile, try spruce chewing gum. Break off a good hunk of the hardened oozing resin and pop it in your mouth. I once chewed a piece for a full day without it breaking down. It's not Wrigley's Spearmint, but it's not bad. Here's one of my favorite recipes. Spruce honey is a delightfully fragrant sweetener that is excellent on wheat toast. In late spring or early summer, gather the new, light green needle growth on the tips of the branches. Rinse the tips and put them in a pan with just enough water to cover. Simmer for about 1 hour. Put the tips in cheesecloth and wring juices out. Add $1/2$ cup of sugar for every 1 cup of liquid. Boil about $1/2$ hour until it thickens, and let cool. Keep refrigerated.

Spruce spirits

Spruce beer was popular with Native Americans and lumbermen in early America. Johann Kohl reported on its use by Lake Superior Ojibwa as early as 1860: "They also refreshed us with a peculiar forest drink, which they honoured with the name of beer . . . The Indians who probably invented it call it, very prosaically, by its right name, 'jingobabo' or 'fir branch water.'" Nineteenth-century philosopher-naturalist-author Henry David Thoreau, in *Maine Woods*, writes about that "lumberer's drink, which would acclimate and naturalize a man at once, which would make him see green, and, if he slept, dream that he heard the wind sough among the pines."

Tamarack
Larix laricina

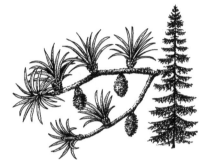

A conical deciduous conifer that loses all its needles in the fall

Short needles in tufts on stem

Needles turn yellow in fall

Grows in wet and boggy areas

Other names

   American larch

   black larch

   red larch

   hackmatack

   muckigwatig ("swamp tree" in Ojibwa)

Trees of the genus *Larix*, Tamarack, Subalpine Larch, and Western Larch, are the only coniferous trees in North America that lose all their needles every year. Technically, they can be called "deciduous conifers." In spring, soon after "the wild geese have gone over and ice in the beaver ponds has melted," twenty to forty delicate bright light green needles burst in tufts from the buds scattered along each stem. Each tree shines in its new cloak of spring green. Watch for this in the first half of May. The inch-long needles stay soft and pliable since there is no need for winter protection. (Most northern conifers need rigid cells and waxy coatings to prevent water loss during the cold months.) By the second and third weeks of October, the North's Tamaracks have turned "smoky gold." What a sight to behold! Especially since they provide such stunning contrast to their "evergreen" neighbors. By early November, the Tamarack has regained its bare-branched military sternness.

Cones are marble-sized and roundish. In a good cone year, twenty thousand may be produced on a single tree.

Minnesota's largest Tamarack grows, not far from my land in Carlton County, at the University of Minnesota's Cloquet Forestry Station. It stands 75 feet tall, has a 40-foot crown spread, and is nearly 9 feet around at chest height. An old Tamarack isn't necessarily a huge Tamarack. Trees enduring life in a low-nutrient bog may be the same age as Minnesota's largest but only 3 feet tall and 1 inch in diameter!

Able to tolerate high acidity, wet soils, and extreme cold, Tamaracks are found in bogs and swamps from Ohio to as far north as the shores of the Mackenzie River and the mouth of the Yukon River in Alaska. Growing to 67° north latitude, it is North America's most northerly tree, even growing by light of the midnight sun at its furthest reaches. Only ground-hugging Arctic willows grow nearer to the North Pole. Within northeast Siberia stand the world's most northerly forests. Here the Dahurian Larch (*Larix dahurica*) survives at 72° north latitude, where January temperatures may plummet to -65° Fahrenheit.

Decay? No way!

Ask good woodworkers what rot-resistant timber they might use and most would say Tamarack or Northern White Cedar. Colonial shipwrights also knew this and sought out sturdy roots bent at angles to use as "knees," which joined the ribs to deck timbers. They sought these out in swamps underlain with hardpan clay, which deflected the roots at sharp angles. In the nineteenth century the Tamarack was used for telegraph poles and railroad ties. Today, fortunately for the Tamarack, green pressure treating can turn any wood into a decay-proof timber.

The Ojibwa of Minnesota gathered the fine rootlets from trees growing on the edges of beaver ponds. These they used to sew together the birchbark pieces that make up a canoe. For burns they applied a finely chopped compress of inner bark.

John Josselyn, the first naturalist-historian of the Bay Colony, claimed that "the turpentine that issueth from the larch tree is singularly good to heal wounds and to draw out the malice of any ache, rubbing the place therewith."

Sparky says: Minnesota's highest scoring Tamarack on the Big Tree Registry is 75 feet tall. To estimate the height of any tree, find a stick (AC) the length of your arm (AB). Hold it upright at arm's length. Move back until you can see the top of the tree when sighting across the top of the stick. Now pace off the distance to the tree (BD). The tree's height will be approximately equal to that distance plus DF, your height, measured to your shoulder.

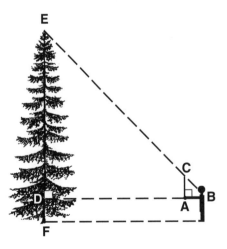

9

Jack Pine
Pinus banksiana

Pioneer species that forms dense mono-stands following fire

Scraggly pine 30 to 70 feet tall

Two stout and twisted needles per bundle

Laden with paired, tightly closed cones

Bark is scaly

Other names

> gigandag (Ojibwa: "gi-GA-n-dag")
>
> scrub pine
>
> banksian pine
>
> black pine
>
> gray pine
>
> black jack pine
>
> princess pine

Jack Pine is a fire specialist, growing in massive mono-stands following wildfires in the North's pinelands. Sealed cones open in the heat and spread millions of seeds in the mineral-rich ash bed left on the charred ground. But fire is not the only difficult condition the pioneering Jack Pine is able to overcome. It thrives in soils that are up to 93 percent sand and essentially as sterile as the local link's sand traps. Wind, heat, sun, cold, and drought are no match for the mighty Jack, but plant one in the shade and just watch it wither. Lower branches can't even tolerate the shade of its own crown. They die and fall off. Bitter cold is no problem either. It is the most northerly pine in North America surviving to 65° north latitude in Canada's Yukon Territory. Local in Nova Scotia and New Brunswick, its range spreads north in the Ungava Peninsula and west nearly to James Bay. The south edge is central Minnesota, central Wisconsin, Michigan, southern Ontario, and New England.

The silhouette of a mature tree is a clear trunk (straight if grown in the company of others or twisted if alone in the open) topped by upswept branches. Thousands of 2-inch long closed cones are visible along the outer branches. Cones start out dull green or purple ripening to pale yellow and eventually gray. They are paired and curl toward the branch. Needles are also bundled by two's. They are about 1 inch long, thick and twisted, with stomata buried deep within the protective outer layer. Needles stay on the tree for two or three years before turning gold and dropping. It is said that a 60 year old, 70 foot tall tree is fast approaching its last days. But in the canoe country they can get much larger and much older. I personally watched forest fire ecologist Dr. Miron "Bud" Heinselman core a big Jack on Minnesota's Seagull Lake. We counted 160 plus rings and dated its origins to the wildfires of 1810! This tree was 180 years old and 2 feet in diameter, and it was not alone!

Fire as friend

Jack Pines owe their life to fire. Fire opens the sealed cones to release the seeds. Without fire, cones can stay as tightly closed as a clenched fist for twenty-five to thirty years, a veritable warehouse of canopy-stored seeds. Viability of the seeds declines after about four years, but most populations spontaneously shed some seeds every year, especially during hot and sunny spells. Two million seeds per acre may be tied up in sealed cones.

The average fire rotation in the boreal hinterlands is every 125 to 180 years—about the life span of a Jack Pine. As a fire sweeps into an area and the heat intensifies to between 115 and 203° Fahrenheit, the resin holding the cones shut melts and the scales separate, releasing the seeds. Cones can withstand heat as high as 900° for thirty seconds and instantaneous blasts of 1292°!

Germination rates are phenomenal in the postfire environment. The postfire seedbed makes a cozy place for seeds locked away from the light of day for two to thirty years. Ground litter has been burned up, so seeds can reach mineral soil. Ground-dwelling granivorous (seed-eating) rodents have also burned up or moved on. Water, sun, and nutrients are now more readily available, and the soil has been warmed up. The only impediments to Jack Pine proliferation are the cones' thick scales, which protect the seeds from squirrels, birds, and intense fire, but which allow fewer seeds per cone, and the long periods between massive seed release, which in this day of Smokey Bear and fire control may never come.

As you paddle or hike or drive through the North Woods, look for the even-age stands of Jacks that signal a fire of yesteryear. With practice you can even age them.

Evil tree

French Canadian woods lore held that a woman's womb would close up tight, like the cones, if she passed within 10 feet of a Jack Pine. Donald Culross Peattie writes in his classic *Natural History of Trees:* "Almost any misfortune that befell a man's ox or his ass or his wife could be blamed on the nearest Jack Pine,

A certain bird's best friend

Kirtland's Warblers need Jack Pines. It's the only tree they'll nest in. And not just any Jack Pine stand will do. No, it must be eighty or more acres of Jacks, all less than twenty years old with their branches reaching down to a dense ground cover. Oh, yes —and no nest-parasitizing Brown-headed Cowbirds allowed. Where do the warblers find such a place? They nest only in the northern lower peninsula of Michigan, with the largest colony around Grayling. U.S. Forest Service naturalists give tours to birders hoping to add this rare warbler to their life lists.

and the only thing to do was to get rid of it. Yet so powerful are the spirits of perversity supposed to inhabit this ill omened tree that no one who valued his life would cut it down. So wood was heaped around it, and the owner then set fire to the kindling. If in its turn it set the tree ablaze, the powers of evil could not blame the man." Of course, we know now that this only increased the cursed trees' abundance.

 Sparky says: Grab some unopened cones off a nearby Jack Pine. Place them on your fire grate or near an open fire. Remember, a temperature of 115º to 203º Fahrenheit melts the resin holding the cones shut. Watch as the scales slowly separate. After they open, snatch the cones from the heat and shake some seeds out. Examine them with a hand lens.

In 1953 Red Pine was voted Minnesota's state tree—a great honor for a tree that had lived in the shadow of its cousin the White Pine for so many years. Historically, Reds were never as abundant as Whites and produced not nearly as desirable lumber. The Red Pine's empire was the north shores of the Great Lakes, including Superior, though it ranged through eastern Canada, New England, south in the mountains to West Virginia, and east to Michigan, Wisconsin, northeast Minnesota, and southeast Manitoba.

Red Pines can grow to humongous proportions during their long lives. The Minnesota record holder stands in Itasca State Park. Two adults joining hands can barely encircle the massive trunk. One hundred and twenty feet tall with a 36-foot crown spread, it scores 245 points on the Big Tree Registry scale. Red Pines appear much darker green and not as "soft" as White Pines. The lower two-thirds of the trunk is usually bare of live branches. Reddish, scaly plates make up the thick protective bark. Needles are 4 to 6½ inches long and bundled by twos. They turn gold and fall in their third October.

Dark purple male flowers grow in dense spikes, while the female flowers form in scarlet bunches at the end of the twig. Flowering happens in May, with the 2-inch long cones ripening in September of the second year. Cones fall the following spring. Bountiful cone and seed crops occur only every four to seven years, but if this does not coincide with good germination conditions, few new trees will sprout. A study in Minnesota's Boundary Waters Canoe Area Wilderness found only one of fifty-six burned Red Pine stands with a good crop of young trees. Once they've taken hold, though, they grow like weeds, about a foot per year for the first sixty years or so. Considering White Pine's problem with Blister Rust, Red Pine is now the most commonly planted pine in the North Woods. In 1962 alone, thirty million seedlings were planted in reforestation projects. But why plant them in sterile, monoculture rows, which are attractive to neither human nor beast? Pine plantations are strangely silent places.

Red Pine
Pinus resinosa

Large, long-needled pine

Needles 4 to 6 inches long and paired in bundles

Bark composed of scaly, reddish plates

Cones small and flexible to 2 inches long

Other names

    Norway pine

    hard pine

    jingwak (Ojibwa: "jing-WAK")

    pin rouge (Quebec)

Fire scars mark many old Red Pines.

The pines of Three Mile Island

On Seagull Lake in the Boundary Waters Canoe Area Wilderness is an island named Three Mile, and it has no nuclear reactor on it. But it does have a stand of Red Pines that fire ecologist Bud Heinselman has dated to 1595! These 400-year-old monarchs bear the scars of many a fire since. Thick, insulating bark keeps ground fires from inflicting much damage to the tree's vital innards. The older they are, the more resistant the trees become. The killing fires are the crown fires that rage through the canopy, destroying everything in their path.

Norway's pine

There is some confusion as to how Red Pine garnered the vernacular name "Norway pine." Some believe that English explorers along the North Atlantic coast mistook the pine for the Norway Spruce of Scandinavia, which was a valuable import for Britain. But how could anybody mistake a spruce for a pine? Maybe they thought it was Scotch Pine, which also grows in Norway and was exported to England. Another theory is that the tree was tagged with the name of Norway, Maine, an important lumber town of the late eighteenth and early nineteenth centuries. But records show that the town was not named until 1797, yet the tree had been called Norway pine as early as 1790. The plot thickens.

Sparky says: Plant some trees. I recently took a nostalgic drive past my junior high alma mater, and I couldn't believe my eyes. The little trees Mr. Beaver's science class planted, watered, and watched were now a one-acre forest! Yes, the trees were big enough to be called a forest. It wasn't that long ago, was it? Well, some were Red Pine, and sixteen years is plenty of time for them to mature. What hit home is that yes, planting trees is an investment in the future, but the rewards can come sooner than you think.

Start by finding a suitable place to plant. Your yard? Your land? A community wildlife refuge? Know what habitat and soil requirements the tree species requires. A Red Pine seedling will not thrive in deep shade or swampy soils. Next, order your stock. Most soil and water conservation districts or natural resources departments will have appropriate species for sale. Special "wildlife packets" are often available. Trees usually come in the "bare-root" form and in bundles of twenty-five or more. They do cost money.

Keep the roots moist until planting. The best tool to use is a planting bar, but lacking this, a shovel will do. Jam the bar or shovel straight down into the earth and pull back on the handle. This creates a wedge-shaped opening, which accepts the roots of the seedling. Use your hand to align the roots so they are pointing straight down. (If your fingernails ain't dirty, you ain't doing it right.) Now take the bar and jam it back into the earth 4 inches behind your first wedge. Pull back first, then push the handle forward until the first wedge is closed shut. Remove the bar and use your heel to close the second wedge. If it's done correctly, there will be no air spaces around the roots and your little tree will be standing vertical. Give it a drink of water. Some trees may need protection from deer, rabbits, and rodents.

Plant a variety of species together and in random cluster patterns. The wildlife will thank you, and I will thank you. There is nothing worse, aesthetically or environmentally, than creating sterile, straight-rowed, monospecific plantations. Planting trees will yield long-term rewards.

White Pine

Pinus strobus

Huge pine with upswept branches

Five needles per bundle

Whorls of five branches

Old bark is ridged and furrowed

4- to 8-inch-long cones with flexible scales

Other names

    Weymouth pine

    soft pine

    pumpkin pine

    sapling pine

    pin blanc (Quebec)

White Pine is most at home in the rocky environs surrounding the Great Lakes. Ranging from Maine to Minnesota, it is not a tree of the far North, penetrating only a short way into Canada. It is, though, the largest conifer east of the Rocky Mountains and the second largest pine in North America. Only the Sugar Pine (*Pinus lambertiana*) of California is larger. Historically, there are accounts of Whites reaching 250 feet tall and 10 feet in diameter. A 240-foot specimen stood on the site that later became Dartmouth College. Minnesota's record holder stands in Itasca State Park. It is 14½ feet around at chest height and 113 feet tall. Eighty to 100 feet is normal adult size. The bark on such behemoths is deeply ridged and thick, protecting the tree from the raging fires that sweep the North Woods. Survivors easily live 200 years, with recorded life spans surpassing 450 years. Branches that sweep up give the tree a "pagoda-like" silhouette.

Five is the magic number. Five delicate needles are in each bundle. They are 3 to 5 inches long. This gives the White Pine's foliage a soft look compared to Jack or Red Pines. White Pine needles are also a brighter, lighter green. Young trees grow fast, 15 to 18 inches per year. Each year a new whorl of five branches grows from the five buds surrounding the central terminal bud at the tree's tip. Finally, at age 20, the pines flower. In certain years pollen can be shed in mass quantities. I witnessed such an event around the Fourth of July, 1993, while standing at an overlook in Minnesota's Boundary Waters Canoe Area Wilderness. With every gust of wind, veritable clouds of pollen were shed. Sunlight was diffused, and lakes were covered in the golden stuff. It coated cameras, glasses, packs, and hair. The result was probably an excellent cone crop, which happens only every three to five years. Cones are 5 to 8 inches long, slender with flexible scales, and hang by ½-inch stalks. They are often tipped with sticky resin.

The king's pine

In 1605 British Royal Navy Captain George Weymouth sailed up the Maine River scouting the New World for Mother England. The crew was in awe of the tallest trees they'd ever seen. Three-hundred- to four-hundred-year-old virgin White Pines towered to 200 or more feet along the banks. Arrow-straight, the boles would make wonderful masts for His Majesty's fleet. Wood-starved England had been piecing together masts from Russian and Swedish Scotch Pine (*Pinus sylvestris*) for years. The ships returned, bringing with them several mast logs and seeds, which were eventually planted at "Longleat," the estate of Viscount Thomas Weymouth. They never did well in the British climate but became known as "Weymouth pines." By 1623 a British-owned mill was in operation at York, Maine. Special mast ships loaded mammoth logs through portals in the bow. The strong light wood was exported to Africa, Spain, Portugal, the West Indies, and eventually Madagascar. In the 1700s settlers began clearing land and torching the Crown's valuable pines. The king became furious and in 1761 declared all White Pines over 24 inches in diameter protected property. Known as the King's "broad-arrow" trees, they were emblazoned with three hatchet marks forming an upward-pointing arrow. The Crown offered rewards to those catching "felon fellers." The reward was the offender's land grant. Settlers took to spying on one another, even going so far as to dress up like the native Indians. Tensions between the colonists and their absentee landlord, England, grew. In 1774 Congress halted all exports to Britain, including White Pine. This helped feed the fires of the coming Revolutionary War. In fact, the White Pine flew on the first flag of the Revolutionary forces, even standing over the Battle of Bunker Hill.

The giants fall

Exploitation of White Pine began with the British in the early

W-H-I-T-E

Pluck a single bundle of needles from any pine in the northeast U.S. Count them. If there are two, you are standing before a Jack or Red Pine. But if you count five, one for every letter in its name (W-H-I-T-E), you are in the presence of a White Pine, the only five-needled pine east of the Rocky Mountains.

1600s, and it continued, almost unabated, for three and a half centuries. The slaughter began in Maine. Lumberjacks working for mills methodically sliced the giants down, working their way west through Ontario, Michigan, Wisconsin, and finally Minnesota by the early 1900s. Their tools were a crosscut saw, horse and wagon, and muscle. Massive pine stands fell, stands that "a squirrel could travel a squirrel's lifetime [through] without coming down from the branches." The immensity of the pine lands is unfathomable to us today. And the size of the virgin timber was of a scale not seen in this age: 200-foot trees proudly standing on trunks 10 feet in diameter. And not just scattered giants but acres of them! The White Pine forest was endless—or so everyone thought. From 1776 to 1940, 2.4 quadrillion (2,400,000,000,000,000) board feet of lumber was cut and milled. Piling all this wood in one city block would create a stack 400 miles high! From this was built 52 million homes, 12 million barns and farm houses, 2 million schools and libraries, 650,000 churches, and 450,000 factories. We are familiar with 2 x 4's, but back then 2 x 36's 15 feet long were common. The wood was the perfect building material, strong yet light and easily worked. It made fantastic timbers. In fact, the building I'm writing in has a superstructure of White Pine beams, 12 inches by 12 inches.

Waste was part of the business. In Minnesota, Michigan, and Wisconsin, over a span of twenty-four years, 85 billion shakes were fashioned from just the prime lower 20 feet of a tree. All the rest (80, 100, 150 feet?) was left to rot. Henry David Thoreau witnessed the carnage and responded by penning these words: "But the pine is no more lumber than man is and to be made into boards and houses is no more its truest and highest use than the truest use of man is to be cut down and made into manure."

White Pine was successfully exploited due to three factors: snow allowed the trees to be skidded easily by horse and wagon, rivers and lakes enabled pines to be floated to mills, and the light weight of pine made it buoyant and manageable. In

Minnesota, between 1839, when the first mill opened, and 1932, when the last one closed, 67 billion board feet of White Pine left the state. But fortunately for canoe country travelers today, many Red and White Pine stands were saved due to the rugged inaccessibility of the rocky land. The Boundary Waters Canoe Area Wilderness and Quetico Provincial Park contain many virgin stands that are protected for future generations.

Bad news blisters

White Pine Blister Rust (*Cronartium ribicola*) is a serious fungal disease that has crippled all efforts at replanting this species. Like most devastating natural pests, it was introduced to America from a foreign ecosystem. This disease arrived around 1908. Young pines are most susceptible, as infections occur on branches within six feet of the ground. But oddly enough, it cannot pass from pine to pine but must have an intermediate host, which in the case of White Pine Blister Rust is any member of the genus *Ribes*, including gooseberries, skunk currants, and currants.

Currants infected with the fungus show hairy rust spots on the undersides of the leaves. It does little harm to the currant, but as the fungal spores are released, they are carried through the air. If they find their way to a White Pine, the spores may eventually find their way into the needle through the breathing pores, or stomata. Once established in the mesophyll tissue inside the needle, the fungus attacks chlorophyll and unmasks the underlying pigment, carotene. The result is a bright yellow spot on the needle. Cool, cloudless, and calm nights leave heavy dews on vegetation, which is conducive to fungal survival and germination. White rootlike mycelial tissues now penetrate the central vascular core and spread down through the needle to the twig and eventually the branch and trunk. The bark becomes yellowed, and grapelike clusters of orange-brown blisters form. They burst, oozing resin down the trunk, leaving telltale white streaks. Inside, the mycelium is rapidly expanding. If it does not kill the tree outright, insects and other fungi will surely finish the job.

All foresters and landowners can do is yank up all currants in the vicinity, prune infected branches before the rust invades the trunk, and trim branches up to six feet above the ground to allow air to circulate, lowering the risk of infection.

In the 1930s Dr. A. J. Riker, a plant pathologist at the University of Wisconsin, started a breeding campaign using seeds of rust-resistant White Pines. (Some of these trees can still be seen in Duluth's Hartley Field.) The Eveleth, Minnesota, U.S. Forest Service nursery continued the work, planting 45,000 seedlings between 1972 and 1974 that were grown from the seeds of 900 rust-free trees. The plot is along the Sawbill Trail, nestled among 7,000 currant bushes. Many will succumb, but those that survive may give hope to the return of the White Pine, monarch of the North.

Pine pals

Mama Black Bears with young cubs prefer to hang out at the bases of large Whites. The thick fissured bark allows the cubs a solid grasp so they can clamber up whenever danger threatens. A Minnesota study showed that moms with cubs chose 16- to 36-inch-diameter White Pines for 90 percent of their beds. Decay-resistant outer wood often results in hollow trees, which are ideal for hibernating bruins.

The open, thick-branched crowns are ideal nesting and perch sites for large-winged birds. A 31-year study on Minnesota's Superior National Forest showed that 81 percent of Bald Eagle nests and 77 percent of Osprey nests were in big Whites, though Whites were only 0.5 percent of all trees. Ravens, American Crows, and Broad-winged and Sharp-shinned Hawks also construct bulky stick nests in large White Pines. The diminutive Pine Warbler also likes the big trees, having been recorded nesting in the tops of fifteen species of pines, including White.

The large seeds are plucked from cones by foraging Red Crossbills, Pine Grosbeaks, Red-breasted Nuthatches, and Black-capped Chickadees. Red Squirrels can strip a cone of all forty-five seeds in about two minutes. In research experiments, radioactively labeled seeds were found to be moved, usually less than 50 feet, by grazing Red-backed Voles and White-footed Mice.

"State-ly" symbols

Maine, Michigan, and Ontario have selected the White Pine as their official tree.

 Sparky says: Very few people, adults and kids alike, can resist stuffing their packs and pockets with pine cones after coming upon a lode while hiking. But then what? They usually sit around the house and get dusty. So why not put them to work as homemade bird feeders? Children love this project. The simplest method is to let them gob peanut butter on a cone (White Pine and Red Pine cones work best) and then roll it in commercial bird seed mix or black oil sunflower seeds. Wrap a wire or string around the base and hang it in a tree that you can see from a window. Chickadees, nuthatches, and woodpeckers will thank you.

Northern White Cedar

Thuja occidentalis

Coniferous tree with flattened scalelike needles

Fibrous bark that sheds vertically

Trunk often curved especially on shores of lakes

Found along lake shores, ringing bogs, and on rocky cliffs

Other names

   arbor vitae

   white cedar

   canoewood

   gijikandug (Ojibwa)

   cèdre (Quebec)

   balai (Quebec)

The Northern White Cedar, as Henry David Thoreau once said, grows where "the woods are all sprucey and moosey." Paddle many northern lakes and you'll encounter the curving trunks of cedars sweeping out over the water. Some lakes and bogs are entirely ringed with them. It is this apparent propensity for water that has given Northern White Cedar the reputation of being a lowland species. But look around. It's everywhere. Hillsides, rocky slopes, and even vertical cliff faces host the arbor vitae. In fact, researchers in Minnesota found that even the cedars in a cedar swamp develop best if the land slopes more than 8 feet per mile. This keeps a good flow of nutrients running through the soil and provides adequate aeration. Cedars prefer a more basic soil pH and therefore are found in limey areas. Fire is another habitat-limiting factor for this thin-barked species. Since cliffs and shorelines are often spared during forest fires this is where the cedars thrive.

Northern White Cedar can take on several shapes depending on the conditions under which it grows. In the woods it can grow straight and conical with branches sweeping the ground (if no deer are present). Along lakeshores and rivers it leans out to catch the sun and so forms arcing trunks, occasionally more horizontal than vertical. Under the harsh realities of life on the rocks, cedars may be scraggly and deformed, mere runts of their woodland siblings. Trunks of all trees taper sharply from base to tip. The thin, fibrous bark sheds in vertical strips and shows a spiraling pattern. Trees spiral to the left when young, then reverse into a right spiral between 75 and 125 years of age. The needles are reduced to flattened scales that overlap one another to form branched "leaves." Look closely at the center leaf scales' underside to find the tiny resin glands that are the source of the cedar's wonderful aroma. Leaves stay on the tree two to five years before turning gold-brown and falling off.

Not a true cedar like the Cedars of Lebanon (*Cedrus libani*), the white cedar is actually a cypress. They are extremely long-lived. Many have been dated at 700-plus years old with one maxing out at over a millennium! Ontario's Quetico Provincial Park boasts a 700-year-old giant at Emerald Lake that has a 4-foot diameter. Minnesota's largest, according to the Big Tree Registry, is 82 feet tall and 11 feet around. It lords over the Little Fork River in Koochiching County (Yes, the home of Rocky and Bullwinkle and Frostbite Falls!). The oldest canoe country tree may be the 4-foot diameter cedar tucked in the woods between Prairie Portage and the Basswood River in Minnesota's Boundary Waters Canoe Area Wilderness. Many paddle nearby but few see it. Its maze of interlaced roots forms a nearly solid mat 28 feet in diameter. A coring revealed 400 growth rings, but the center was punky. Extrapolating the number of rings needed to fill the inner core, an age of 1,100 years was estimated!

Chronologically gifted

The largest stand of old growth cedars in eastern North America was not discovered until 1988. It happened when University of Guelph botanist Doug Larson was exploring the Niagara Escarpment, a 465-mile-long cliff face running north from Niagara Falls to Tobermony, Ontario. The 80- to 100-foot vertical cliffs are composed of dolomitic limestone. On the face cling thousands of stunted Northern White Cedars. Enduring strong winds and soil temperatures of -13° Fahrenheit, the gnarled cedars have managed to survive their precarious position for, in many cases, 700 or more years. Counting rings on such small and old trees required the use of a microscope. One 652-year-old specimen had the trunk diameter of an American silver dollar! Under normal conditions, a cedar of that size would only be five to ten years old. The oldest tree confirmed on the escarpment was a 1,032-year-old cedar. And that is actual rings counted, not estimated. To put this in perspective, that cedar was already over 500 years old when Columbus landed in America! Researchers found that the oldest trees were growing in microhabitats that were at either the driest and/or coldest end of the spectrum. No one knows why. But it is true that these growth rates are the lowest yet recorded in the plant

Nursery logs

Fallen Northern White Cedars covered with moss provide a safe haven and sprouting bed for seeds of its own kind. These are called "nursery logs." They hold water better than soil and provide a haven from allelopathic (poison) compounds secreted by Sugar Maples. In one study, such fallen giants, though only covering 16 percent of the forest floor, supported 75 to 85 percent of all Northern White Cedar seedlings present.

kingdom. Some Niagara Escarpment cedars are only adding 0.9 grams in weight each year, an amazing 1,000 percent less than published values. Botanists Doug Larson and P. E. Kelly discovered one 530-year-old cedar that, minus foliage, weighed a scant 140 grams (5 ounces!). This calculates out to an infinitesimal 0.26 grams (less than one ounce) of wood weight added per year. The majority of annual rings were only two or three cells thick! Though not as visually awesome as the California sequoias (*Sequoya* species) nor as old as the Bristlecone Pines (*Pinus aristata*), the Northern White Cedars of the Niagara Escarpment are nonetheless an amazing story of endurance and survival.

Tree of life

Jacques Cartier, the famous French explorer, and his crew were saved from the agonies of scurvy by the Northern White Cedar. On their voyage up the St. Lawrence River in 1535–36, they fortuitously encountered the white cedar and an Indian named Domagaia. Just ten days earlier, Domagaia had been near death, "very sicke and . . . his knees swolne as bigge as a child of two years old." A tea of bark and leaves boiled together and drunk every other day cured him and saved Cartier's men. This miracle tree was christened *arbor vitae*, the tree of life in Latin. We know today that the needles are very high in vitamin C.

An old lumberjack tune calls for "A quart of arbor vitae to make him strong and mighty!" This song was based on the belief that a cedar tea prevented rheumatism. Euell Gibbons, the late wild edibles guru, ranted that "after trying this tea in several strengths, with and without sugar and milk, I've decided I would almost prefer rheumatism."

The Menominee women made a tea from the inner bark to promote menstruation. Their neighbors the Ojibwa took cedar needles and wood from the Ironwood (*Ostrya virginiana*) to make a cure for coughs. Scientific research has shown that oil of cedar is a stimulant to the bronchial passages, uterus, and other smooth muscles. Excessive phlegm associated with bronchitis is treated by modern herbalists using the oil. An ointment is reputed to cure warts. Pain from rheumatism is eased with a hot compress. Infectious skin diseases such as scabies and impetigo are alleviated with an infusion. But caution is advised, as oil of cedar is comprised of camphor and 65 percent thujone, a heart stimulant that is toxic in any decent dose. Consult an expert. And NEVER take oil of cedar during pregnancy!

Uses for cedar

Highly rot-resistant, white cedar is an excellent choice for situations where water comes in contact with the wood, for example, fenceposts, shakes and shingles, canoes, paddles, fishing floats, and lures. In the days before creosote, cedar was also used for railroad ties and telephone poles. The wood is very light, a mere 29 pounds per cubic foot and soft (Ironwood, by contrast, is 51 pounds per cubic foot). It doesn't shrink or warp. Cedar is too brittle to make a good beam, yet a shaving is flexible enough to bend in two and not break. The Ojibwa took advantage of this trait by weaving floor mats out of the inner bark in spring.

Domesticated

Northern White Cedar was most likely the first North American tree to be exported and cultivated in Europe. It was established in Paris between 1536 and 1550. Today there are over fifty ornamental varieties of arbor vitae available to the homeowner for incorporation into a landscape design.

 Sparky says: Are fallen trees and stumps worth anything? Take a close look at one that has become a "nursery log." What types of plants have found a home there? There are probably mosses. How about flowers? Lichens? Fungi? Can you find any seedling trees of the same species as the nursery log? Make a list of all the organisms utilizing such an apparently dead and useless thing as a fallen log. Do you have a new opinion about logs and stumps?

Wild associates

Red Squirrels nip off cone-bearing branchlets, storing the cones for winter. Pine Siskins are the primary foragers of cone seeds. Swainson's Thrushes use fine strips of the bark to line their cup nests. A favorite food of White-tailed Deer, munched cedars create a sharply defined browse line visible along lakeshores. Dry cedar uplands are the best places to find the elusive Calypso Orchid (*Calypso bulbosa*). Search them out in late May and early June.

Calypso Orchids can be found under dry cedar stands.

Paper Birch
Betula papyrifera

White-barked tree

Peeling, papery bark is marked with bold black horizontal lenticels

Leaves oval, pointed, and toothed

Male and female catkins hang on the same tree just before leaves unfurl

Other names

    canoe birch

    white birch

    wigwasatig (Ojibwa: "WIG-was-a-tig")

    bouleau blanc (Quebec)

No deciduous tree is more symbolic of the North than the Paper Birch. All the northeast's Indian tribes made use of the waterproof paper bark for containers and for the essential mode of early transportation, the canoe. Legends from native peoples on both sides of the ocean attest to its enduring place in their cultures. Even the Sammi (Lapplanders) of Scandinavia, Finland, and Russia worked the bark of this circumpolar genus into functional tools and works of art. It is a tree that can't stand the heat, liking cool and moist weather. July and August temperature averages of 70º Fahrenheit are just too much for this pale-skinned northerner.

Paper Birch is an early successional species, growing in the company of Quaking Aspen only to be overtaken years later by the climax species Balsam Fir and White Spruce. It grows in two ways. A clump of two to six trunks, all arising from a common point, indicates that these birches started life as sprouts from the base of a long-gone parent tree. If the clump of trees is young, the rotted stump may still be present. Other birches grow as single-trunked trees. Most likely these originate from wind-dispersed seeds. The Paper Birch's life span is relatively short. An 80-year-old tree has seen its best days. Since the bark is waterproof, the wood rots from the inside and the tree becomes susceptible to blowdown. Pick up a fallen section sometime and shake the punky wood out. You are left with a hollow tube of intact birchbark. I'm sure this is how the native inventor of the first canoe or basket discovered the amazing water repellency and flexibility of birchbark.

Young trees, at least for their first decade or so, are wrapped in a smooth rich-brown bark that is speckled with lighter horizontal lenticels. Gradually this peels away to reveal the familiar ghost-white papery bark. As the tree grows, the outer layer splits and peels to reveal a new layer beneath it. The trunk is distinctly marked with large black "eyes" that are raised and rough. These are actually winged branch scars. The smaller, more numerous black streaks are lenticels that allow exchange of gases (including oxygen) between the tree's inner tissues and the outside air.

Leaves are heavily toothed, oval, and pointed, 2 to 4 inches long. Undersides are hairy in the vein angles. In a good fall color year, leaves turn a glowing gold, peaking in the canoe country from late September to early October. Male and female flowers are both borne in catkins that developed the previous summer but descend just before leaf-out in mid-May. Female catkins are short (1½ inches) and erect. Male catkins are longer (3 to 4 inches) and dangle from stem tips. Spring winds spread the 5,500,000 pollen grains per catkin far and wide. If pollinated, the female flower matures into a receptacle for many winged seeds (shaped like the Girl Scout crest, says one friend). Gradually disintegrating through the fall and winter and spread by the wind, look for them on the snow beneath birches.

Barking up the right tree

No other living thing was as intertwined with Ojibwa culture of yesteryear as the bark of the Paper Birch. Large sheets of it were harvested in the spring and used to cover the sapling frame of the domed summer dwelling, the wigwam. Rolled up and carried with the Ojibwa, birchbark sheets were also used to cover the conical, tipilike winter shelters. The birchbark canoe was the quintessential form of transportation along the watery highways of the North. Bark sheets sewn together with bast of spruce root were given shape by cedar ribs, gunwales, and thwarts. So accomplished were the native craftspeople that the great fur companies of the 1700s, the North West Company and the Hudson's Bay Company, hired them to make the massive 36-foot-long *canoe d'maitre*, which floated four tons of goods and required twelve to sixteen voyageurs to paddle!

Ojibwa women made narrow-mouthed, fat-bottomed *makuks* for gathering and storing berries and maple sugar. Moose calls, trays, funnels, torches, tinder, rice-winnowing pans, patterns for quillwork, and coverings for the dead were all made from the white skin of the birch. The unique decorative art called ojibagonsigen utilized a very thin single layer of bark cut in a square and repeatedly folded. The woman then bit a

Sweet as honey

Birches can be tapped for syrup just like Sugar Maples. Of course, the sugar content of the sap isn't as high, so you have to boil it twice as long. Russians and Scandinavians let the sap ferment to make wine and vinegar. Finns drink the sap straight or concentrate it a bit by repeated removal of skim ice. Birch beer can also be manufactured from the sap.

Birchbark canoes provided transportation throughout the North.

Useful for small items

Birch is one of the few hard-woods in the North (39 pounds per cubic foot). It makes a good firewood. Toothpicks, Popsicle sticks, clothespins, broom handles, and those relics of the past, wooden spools, were all at one time or another manufactured from Paper Birch.

pattern into it that, when the square was unfolded, formed an intricate floral or geometric design.

The Sammi of Lappland also made baskets and trays out of birchbark. But they wove the strips of bark, unlike the Ojibwa, who used whole pieces. Was this the result of smaller, more stunted trees in Lappland or just cultural preference?

Why are they dying?

Birches are prematurely dying off in huge numbers throughout the canoe country. Why? Foresters believe that the drought and heat of the years surrounding 1988 weakened the trees dramatically. With the trees' resistance down, the Bronze Birch Borer (*Agrilus anxius*) invaded, the larva of this buprestid beetle girdling beneath the bark and eventually killing the tree.

Sapsuckers and siskins

Birch seeds hanging in catkins provide an important midwinter food for flocks of Pine Siskins, Common Redpolls, and Black-capped Chickadees. Young twigs make favored browse for Snowshoe Hares, cottontails, Moose, White-tailed Deer, and Porcupines. Porkies also munch on leaves. Philadelphia Vireos and Black-throated Green Warblers use fine strips of bark to decorate the exterior of their nests. Yellow-bellied Sapsuckers drill neat rows of holes in the bark in order to lick up the sweet sap. On the next round of their "sap trees," the sapsuckers may also be rewarded with protein-rich insects. Hairy and Downy Woodpeckers, nuthatches, and Black-and-white Warblers find the peeling bark to be a good source of insects and their eggs. Beaver enjoy the inner bark, and Ruffed Grouse gorge themselves on winter buds.

Nanaboujou

An Ojibwa legend tells how the birch got its black markings. Nanaboujou, the trickster, coveted the powers of the thunderbirds. So he snuck up to their nest high on the cliffs above Lake Superior and transformed himself into the shape of a nestling. The ruse went on for several days until he was discovered. He fled to earth, transmuting to the shape of a rabbit on the way. The thunderbirds were on his tail—literally, when he spied a hollow birch log. In he ran. The thunderbirds, unable to stop, slammed into the trunk, leaving for all time the impression of their body and wings on the bark.

 Sparky says: Find a piece of downed birchbark or strip some off a fallen tree. Cut a thick piece to the shape of a postcard. Jot a note to someone back home, put a stamp on it, and mail it.

If you get ambitious, try a basket. You can make a shallow rice-winnowing tray by folding the corners in on a rectangular piece of bark and then sewing a willow rim to it. If you have no split spruce root to sew with, try rattan, which can be found at most craft stores.

If you get bored, have a leaf-flying contest. Everyone chooses a leaf they think is a good flyer. Stand behind a line and release the leaves. The person whose leaf travels the furthest wins. Often a gentle, floating release is better than an actual throw.

Quaking Aspen

Populus tremuloides

Fast-growing, pioneer deciduous tree

Petiole longer than leaf

Leaves shimmer in slightest breeze

Bark is smooth and whitish when young

Other names

 trembling aspen

 popple or popal

 poplar

 small-tooth aspen

Quaking Aspens, known to my neighbors as popal or popple, have probably the widest range of any native North American tree. They dot the continent from the treeline in Canada and Alaska south to northern Mexico. Aspens are true pioneers, invading burned, logged, or mowed sites. Extremely fast growing, they reproduce by root suckering as clones of the parent tree. Young bark is smooth and greenish, becoming a chalky white or cream. Claw marks and carved graffiti become blackened scars for years to come. Old tree trunks are dark and furrowed, and if you didn't gaze upward, you'd think you were at the base of an old White Pine.

Saplings first flower around their fifteenth year. The male and female catkins are on separate trees and descend about two to three weeks before the leaves appear. In northern Minnesota, flowering takes place in mid to late April. Spring winds accomplish pollination. Leaves unfurl in the middle of May. Toothed triangular-shaped leaves are borne on petioles (leaf stalks) that are longer than the leaf itself, up to $4\frac{1}{2}$ inches long. Quaking, or trembling, aspen acquired its name from the "twinkling" foliage that seems to shiver under the slightest breath of wind. The long, flattened petiole is at a right angle to the leaf blade and therefore flutters the leaf, alternately revealing the shining green top and the silvery underside. Short-lived, aspens begin deteriorating at thirty-five to forty-five years, but may live to eighty. I don't know how old Minnesota's largest is, but—if it's still standing—it's 101 inches around at chest height and 112 feet tall and has a crown spanning 38 feet.

Clones

Quaking Aspens can produce offspring genetically identical to the parent tree through root suckers. If the aspen parent is a female, for example, all trees in the clone (all trees genetically identical to parent) will produce only female flowers. Below Duluth's Hawk Ridge, one clone has always held its golden autumn leaves a full week longer than any other Quaking Aspens in the area. Evidently its parent had a special leaf-retention gene.

Since adult trees produce hormones that inhibit production of suckers, only after fire or logging, when both shade and growth regulators are gone, do the root suckers spurt up. And I mean spurt. In full light an aspen sucker can grow 3 feet per year for ten years! They can attain this phenomenal growth rate because they arise from an already established root system, full of stored nutrients and water. One year after disturbance, an incredible 40,000 suckers may cover one acre. But only the fastest growing survive, shading out their neighbors in the process. After twenty-three years, 95 percent have died, leaving an even-height stand of 2,200 trees. In the Lakes states, individual clones usually cover less than an acre, though clones comprising 50,000 trees on two hundred acres have been found. The root systems that support clones may be the oldest living things on the planet. One in Minnesota was aged at 8,000 years! This makes California's 4,600-year-old Bristlecone Pines mere infants.

Weed tree

Popals were at one time, in the not-so-distant past, disdained by the timber industry as "weed trees." The weak, brittle wood did not make good lumber or firewood, but practically took over an area once it was logged. What to do with such an abundant, yet useless, tree? The revolution came in the 1970s, when experiments proved that chipped aspen, glued together to make boards, could be an inexpensive and feasible lumber product. Also, with the burgeoning information society, the paper and pulp industry took off. All of a sudden the lowly aspen was in the limelight. By 1981 half of all the timber cut in Minnesota was aspen. Superwood of Duluth makes hardboard. The Boise Cascade plant in International Falls makes reams of photocopier paper. Grand Rapids' Blandin specializes in magazine and catalog paper, while Potlatch of Cloquet produces paper for advertising brochures, corporate annual reports, and envelopes. But popal's popularity is not limited to paper and chipboard. A successful chopstick factory operated on

Photosynthetic bark

As one of the most northerly deciduous trees on the globe, Quaking Aspens face cold and dark conditions that evergreens are superbly created for. Thicker, whiter bark on the south side of trunks reflects heat and prevents injurious thawing in midwinter. The bark is also photosynthetic, extending its food-making season on both ends and adding significantly to the annual carbohydrate gain of the tree. The bark reutilizes 50 to 75 percent of respiration-released carbon dioxide in photosynthesis. It all helps, because deciduous trees constantly lose carbohydrates via respiration throughout the winter.

Bugs and bears and birds

Over three hundred species of insects feed on aspens. The Big Poplar Sphinx Moth (*Pachysphinx modest*) starts its life as a caterpillar feeding on aspen leaves. One acre of Quaking Aspen could support a colony of five beavers for three years. Porcupines also enjoy the bark, occasionally nipping off entire twigs and eating just the leaves, leaving the petioles in place. In late April I've seen porkies downing newly de-scended catkins like candy. Black Bears find the newly emerged leaves to be a great after-hibernation snack. Ruffed Grouse are intimately tied to Quaking Aspen. They gorge on the male flower buds in winter, nest at the base of mature trees, and use tangles of saplings for protection in summer.

Minnesota's Iron Range for several years, pumping out the Asian-style utensils four times faster than the Japanese themselves. Farmers today are testing a new cattle feed composed of aspen bark and wood residue. And now, of course, the "weed tree" of the past can't grow fast enough for the massive mills. Hybrids, developed to mature rapidly, are pushing skyward as we speak.

Aspen enemy number one

What do 1922, 1937, 1952, 1967, 1978, and 1989 have in common? If you said years in which Forest Tent Caterpillars (*Malacosoma disstria*) decimated Minnesota's aspens, you'd be right. The mistakenly named "army worms" appear just as leaves begin to open. Fueled by aspen leaves, the larvae grow through five stages, maturing into 2-inch caterpillars striped with blue and marked down the back with cream-colored "footprints." In bad years defoliation is complete, and the "pitter-patter" of frass (caterpillar poop) sounds like rain on the forest floor. Four million may inhabit a single acre! Houses are covered and roads become slippery as the caterpillars move in search of food. Panic-stricken citizens coat the trunks of orna-mental trees with aluminum foil and cooking oil. But in the end, an annoying large gray fly (*Sarcophaga aldrichi*) with rust-colored eyes comes to the rescue. Females lay eggs on the tent caterpillar cocoons. The maggots invade the silk and kill the pupae. Summer temperatures over 100° and winter cold below -42° also kill many. Starvation also wreaks havoc on their numbers.

If the cocoon escapes parasitization, the beautiful caterpil-lars metamorphose into rather dull brown moths. They mate, and the female lays clusters of 100 to 350 eggs that encircle 1- to 2-year-old twigs. Cemented together, the mass hardens into a glossy brown case. Eggs overwinter and hatch the following spring.

But the destruction is not as bad as it appears. Even com-pletely defoliated Quaking Aspens will grow new leaves before the summer is through, albeit smaller and clustered at branch tips. Invasions last three years with one year noticeably worse than the others. Duluth will be due sometime just after the turn of the twenty-first century.

 Sparky says: Trees are living organisms. This activity, called the Tree Machine, is designed to graphically illustrate that point. All you need is fifteen to thirty children or gung-ho adults, who will be arranged into concentric rings simulating the structure of a tree. As you go, explain the functions of each tree feature.

Start with one or two people to be the heartwood, providing support for the tree. They simulate a heartbeat by chanting "da-dub, da-dub." Around them place people to be the xylem. Explain that the xylem transports water and dissolved minerals up the tree. Their backs should be to the heartwood and they should all hold hands, which they raise and lower in unison while droning "xy-lem, xy-lem." The cambium is the growing part of the tree. The next group encircles the xylem, holds hands, and repeats the word "grow, grow." Phloem moves fluids and dissolved food down the tree. This ring of people pumps their joined hands up and down, saying "phloem, phloem." The last ring of humanity plays the role of bark, protecting the tree from disease and injury. They repeat in unison "protect, protect." Now, sprawl a few leftover people on the ground radiating out from the tree. These are the root hairs that absorb water and minerals from the soil. Get them to make slurping noises. You now have a working, living tree. Over the noise, tell the participants that you are an invading insect or fungus spore and you're going to try to get to the core heartwood. It won't be easy!

Pin Cherry

Prunus pensylvanica

Small tree of disturbed areas

Numerous five-petaled white flowers

Blooms from mid-May to early June

Fruits are partially translucent and glowing red

Cherries ripe in July

Other names

baewiminun (Ojibwa: "BAE-wim-i-NUN")

fire cherry

bird cherry

pigeon cherry

wild red cherry

This small tree is especially common in rocky areas following forest fires. Usually under 20 feet in the North, it can attain mutant sizes in the Great Smoky Mountains (30 to 40 feet tall and 20 inches in diameter). The bark of young trees is a deep red-brown flecked with lighter horizontal lenticels. Smooth and papery, it peels in strips. Like all members of the rose family, the flowers are five-petaled and many-stamened; the pin cherry's are white. Clusters of four or five hang by 1-inch stems originating from a common point, like the struts of an umbrella. They appear when the leaves are half grown, in late May or early June. Fully developed leaves are lance-shaped, finely toothed, and 3 to 4 inches long. Cherries ripen in July and August. Partially translucent, the small, red cherries seem to glow in the sun. A large pit is found in the middle. Birds are wild about the cherries, which, unfortunately for people but fortunately for the birds, are too sour for human enjoyment.

Bird cherry

Twenty-three species of birds are known to pig out on Pin Cherries. Grosbeaks, waxwings, robins, and other thrushes are frequent feeders. One observer found forty-seven species feeding on three species of *Prunus*. Another researcher compiled a list of fifteen mammal species that fed on the fruits of *Prunus*, which also includes Wild Plum (*P. americana*), Sand Cherry (*P. pumila*), Black Cherry (*P. serotina*), and Chokecherry (*P. virginiana*).

In 1890, an ornithologist imprisoned a young Cedar Waxwing for twelve days. The parents made an incredible 140 visits per day, bringing an average of five cherries each time (one in the beak and four in the throat). Total number of Pin Cherries fed in twelve days: 8400!

Pin Cherries grow fast and die young. Twenty-five to thirty years is all they last. During their lifetime, thousands of fruits drop to the ground, where the seeds lie in the shade, an environment nonconducive for germination of a sun-loving tree. This is why you'll never find Pin Cherry trees in any old field successional stage. They need the parent trees to provide the original seed source. The fugitive Pin Cherry seeds (aka pits or stones) may remain viable on the forest floor for fifty years or more. A coating impermeable to water or oxygen prevents them from rotting. Two wooded sites in New Hampshire, nearly devoid of healthy Pin Cherry trees, revealed an unbelievable 140,000 to 200,000 dormant cherry seeds per acre! All just waiting for a disturbance—windfall, fire, logging operation—to open up the canopy and let the sun in. Once such an event happens they germinate en masse, creating a nearly solid stand of even-aged trees. Such plants are called "fugitive" species.

 Sparky says: Pucker up and sample some sour Pin Cherries. Now you can really appreciate the sweetness of good Pin Cherry Jelly! You'll need:

6^1/$_2$ cups Pin Cherry juice
7 cups sugar
1 box powdered pectin

Crush cherries, add water, and simmer for 10 minutes. Strain juice through cheesecloth and measure. Now add juice to sugar and pectin. Bring to a rolling boil for 1 full minute, being sure to stir constantly. Remove from heat. Skim off foam and pour into sterilized jars. Cover with sterilized lids. Voila!

Juneberries

Amelanchier species

Small tree or shrub found on many sites

Spectacular white bloom before leaves appear

Flower has five long delicate petals and numerous stamens

Blooms May 10 to May 30

Fruits turn from red to a deep purple

Berries ripen in July

Other names

serviceberry

shadbush

shadberry

makwimin (Ojibwa)

In the North Woods, juneberries (or serviceberries, as many know them) rarely become more than bushes or small trees, possibly reaching 10 to 20 feet. In the south they can grow to 60 feet tall, with a sturdy trunk 2 feet in diameter. All have smooth gray bark that is lightly streaked longitudinally. Juneberries are very difficult to separate in the field. There are over a dozen closely related species. Twelve species have been recorded in Minnesota's Boundary Waters Canoe Area Wilderness alone. Identification is further complicated by subspecies and interspecific hybrids. Low Juneberry (*Amelanchier humilis*), a thicket-forming species, is common on shorelines and in bedrock crevices of cliffs. Northern Juneberry (*A. bartramiana*) can stand the acidic conditions of sphagnum bog life. Saskatoon (*A. alnifolia*) is the very large-fruited species that is also common on the western prairies.

All put on a spectacular spring show. The small leafless trees loaded with white blossoms jump out of the landscape. As a member of the rose family clan, juneberry flowers have five petals and many stamens. The petals are long, slender, and delicate, four times longer than the sepals. The show in northern Minnesota usually begins about fishing opener. May 10 to May 30 provides the best time to look. The leaves follow shortly thereafter. Oval 3-inch leaves are rounded at both ends with just a bit of a point at the tip. They are finely toothed, but this feature is not visible from any distance. The fruit ripens from red to deep purple—it's not ripe until it attains this rich hue. Don't be fooled, or your taste buds will be disappointed. As with the blueberries, dried remnants of the flower parts stay attached to the bottom of the fruit.

In the North they should be called "Julyberries" for that's when they ripen. Juneberry trees often spread by underground stems.

Lewis and Clark lifesaver

Berries of the *Amelanchier* species, Saskatoon, were mixed with pounded and dried buffalo meat and fat to make pemmican, an

Indian specialty and a staple of the Lewis and Clark expedition. The Indians also made the large purple berries into dried cakes, which they broke up and cooked with meat and vegetables.

 Sparky says: My first juneberry pie was an eye-opening experience. It was fantastic! Every bit as good as a blueberry pie. The fruits stay solid, but the seeds get quite soft and are barely noticeable. And this from trees that grew throughout my neighborhood. Use your favorite blueberry pie recipe and substitute fresh wild juneberries. Also try them plain, in muffins and pancakes, as jam, or with cream and sugar.

American Mountain-Ash

Sorbus americana

Small tree with eleven to seventeen leaflets on each pinnately compound leaf

Grows on rock ledges, lakeshores, low rich woods, and even sphagnum bogs

Clusters of creamy white flowers bloom throughout June

Red-orange berries ripen in August

Other names

   American rowan tree

   mountain sumac

   wine tree

   pihala (Finnish-American)

   mikomij (Ojibwa: "MI-ko-MIJ")

   Indian mozemize

   "moose-miss"

   "missey-moosey"

American Mountain-Ash and the closely related Northern Mountain-Ash (*Sorbus decora*) stand out in three seasons. In late spring to early summer clusters of creamy white flowers adorn the dark green foliage. Like all members of the rose family, the flowers are five-petaled with many stamens, but are so small that they lose their identity to the gaudy cluster itself. Then in August, the pea-sized, berrylike pomes ripen to a rich red-orange. The leaves, which turn gold in September, add to the striking combination. Winter trees are hardly dreary, bearing numerous clusters of festive red berries. They literally come alive when flocks of wandering Bohemian Waxwings descend to their fruit-laden branches. City cousin to the native mountain-ashes is the domestic European Mountain-Ash (*Sorbus aucuparia*). Common in the urban landscape, it has rounder leaflets and oranger fruits.

A small tree, reaching 30 feet in rare instances, American Mountain-Ash grows in a wide variety of edge habitats. In the canoe country, it is very conspicuous as it clings to rock ledges and shorelines. But it also can thrive in rich woodland soils and even in low-nutrient sphagnum bog margins with Black Spruce and Speckled Alder. Young bark is thin and papery, a rich shining brown marked with light horizontal lenticels. Six- to 8-inch-long leaves are pinnately compound, which means that a single leaf is composed of eleven to seventeen 2- to 4-inch-long, finely toothed leaflets along a central petiole. Birds who dine on the fruits help distribute the tree far and wide. The fernlike seedlings are very common on the forest floor in certain regions.

Witchwood

Celtic Druids practically worshiped the European Mountain-Ash, or rowan tree, during the fifteenth and sixteenth centuries. This reverence by alleged wizards and sorcerers led to its reputation as an evil tree, or "witchwood." In England it became a symbol of paganism and the supernatural. Ironically, in Scotland the rowan tree was planted next to every cottage front door to keep witches at bay.

Fruit for thought

Are mountain-ash fruits edible? Yes and no. Fresh fruits reportedly contain parascorbic acid, a cancer-causing compound. There is some evidence that they may be toxic to kids. But fortunately they taste quite bitter, and nobody in their right mind would eat too many. A good hard frost and cooking will cut the astringency and make the fruits quite edible. Since they are high in pectin, mountain-ash berries make excellent jelly. The fruits are high in vitamin C and ascorbic acid. Ascorbic means "without scurvy"; herbalists once prescribed mountain-ash fruit tea as a cure or preventative for scurvy. This acid is also an excellent rectal wash and therefore is still suggested for diarrhea and hemorrhoids.

Low-fat fuel

Though the red-orange berries are low in fat, they are extremely attractive to birds due to their abundance in a season when food is scarce. Thrushes, Evening Grosbeaks, and Pine Grosbeaks all imbibe, but the waxwings are veritable fruit vacuums. I remember one cold December day in Grand Marais, Minnesota, when the air was full of the high, thin whistles of Bohemian Waxwings. In every European Mountain-Ash and crab apple tree, one hundred to two hundred waxwings busily downed berry after berry. I estimated five thousand Bohemians in an area of ten city blocks. The town was stripped clean in three or four days, and the waxwings moved on to their next mountain-ash mecca.

Moose relish the fragrant inner bark, foliage, and twigs of mountain-ash. Yellow-bellied Sapsuckers drill rows of holes in the bark to gain access to the sap and the insects attracted to the sap.

Bohemian Waxwings thrive on mountain-ash berries.

Sparky says: Here's a fun family matching game that will help younger kids learn about their natural surroundings. Gather up a dozen or so items with unique shapes, textures, or colors—maybe a maple leaf, piece of fallen birchbark, or a colored rock. Give one item to a child and have them lead you to their match. Talk about how they found it, what it is, and why it's that way. Make it a game, not a contest, so everyone wins.

Most northerly of all the ashes, Black Ash keeps company with Northern White Cedar, Black Spruce, and Tamarack in the low, wet areas. They thrive in the rich, wet clay or silt soils found in swamps and along rivers and streams. Black Ash trees can grow to monstrous size, but the majority die before attaining such proportions. Minnesota's largest is a Koochiching County specimen 101 feet tall, 111 inches in circumference, topped by a crown spreading 52 feet.

Probably the last tree in the North Woods to leaf out in the spring, Black Ash sprouts 13-inch compound leaves composed of seven to eleven finely toothed leaflets, each 4 to 5 inches long. But the last to get leaves is also the first to lose them. Even before summer is officially over, the leaves have yellowed and begun dropping. Because the Black Ash stands bare of leaves eight months of the year, you should learn to identify its unique silhouette. Lower branches droop, but each successive tier of branches is held higher and higher until the uppermost, which bow upward. If ever in doubt, check the outermost twigs, which are very fat. Up close, the tan twigs and chocolate brown buds resemble a miniature deer leg and hoof.

Even before the leaves appear, the odd flowers burst from their black, covering scales. "Odd" because some are unisexual and others "perfect," bearing both male and female parts. Female flowers ripen into a unique fruit form called the samara. A samara has one wing and looks like half of the familiar maple seed "helicopters" we all played with as kids. These 1 1/2-inch-long samaras hang in open bunches 8 to 10 inches long. On more than one occasion I have witnessed Pine Grosbeaks making a winter meal of them.

Knock on wood

The Ojibwa took knocking on wood to its highest form with Black Ash. After peeling and soaking the log, they would beat it from top to bottom with a birch mallet. The spring wood, the portion of the annual growth ring formed when growth is speediest, is composed mainly of large pores. When pounded,

Black Ash
Fraxinus nigra

Large tree of the wet areas

Last tree to leaf out in spring and first to lose leaves in September

Lower branches droop while upper branches arch upward

Leaves 13 inches long, comprising seven to eleven leaflets

Fruit is a one-winged achene called a samara

Other names

    swamp ash

    agimak (Ojibwa for "snowshoe tree")

    water ash

    basket ash

    hoop ash

    brown ash

    frêne noir (Quebec)

Ash was essential for snowshoe frames, and snowshoes were essential for winter travel.

this layer easily separates from between the denser summer wood layers into long strips of even thickness. The Ojibwa wove these splints into baskets of many kinds.

They also made lacrosse sticks and snowshoe frames of Black Ash. In fact, so crucial was ash for the frames that snowshoes were called *agim* and the tree *agimak*. Every part of the shoe had a name. *Oshkinjig* refers to the "eye" of the snowshoe where the foot pivots through. The master cord the foot pivots on is *bimikibison*, and the cross braces are *okanik*. Even the ornamental red tassels (*nimaigan*) and binding's heel band (*adiman*) had specific names. Lacing was made of rawhide. The Ojibwa-style snowshoe is still popular today. A pointed tail, as well as a pointed and upturned toe, slashes through heavy brush, submarines up through crusty snow, and tracks well on open lakes.

Ash birds: the birder's nemesis

I don't know how many times I've slowed the car to check out an interesting bird perched high in a leafless ash tree, only to have it transmute into a clump of dried black junk through my binoculars. What was this stuff? It seemed to be only in Black Ash trees. It wasn't dead leaves or seeds, I knew that. A little research revealed that these mysterious "ash birds" are really aggregates of malformed male flowers misshapen by disease or insects. This eases my mind a bit, but I know next winter I'll be slamming on the brakes to have another good close look at some "ash birds."

Early American uses

George Washington, while surveying his land along the Kanawha and Ohio Rivers, notched several "hoop trees" to serve as property line markers. We know that these were Black Ash, which during that era were commonly used as hoops, or bands, around oaken barrels. Washboards and church pews were also crafted out of Black Ash. Though closely related,

Black Ash does not exhibit the strength and shock resistance of White Ash, which has been used to make baseball bats for over one hundred years. A Black Ash bat would disintegrate into a splintered piece of kindling by the seventh-inning stretch!

 Sparky says: The Ojibwa wove baskets of split Black Ash and you can too. Start with a knot-free 6-foot section of Black Ash. Ideally the log should be 12 inches in diameter and should be felled in the spring when the sap is up. Remove the bark immediately. Now pound down the length of the log with a heavy wooden mallet or club. You are crushing the layer between annual rings, causing them to separate. Gently and steadily pull the loose splint as you go, continuing to pound every square inch of wood ahead until it is removed. Rotate log and repeat around entire circumference. You'll find that splints peel off quite uniformly but may need to be scraped smooth with a paint scraper. Once you've accumulated a pile to work with, it's time to start weaving. If splints have dried out, soak them before beginning. Plait a simple over-under weave to give the desired bottom size. Now bend the sides up to a 90° angle over a table edge. The first weaver of the side wall should be held in place with a clothespin. Hide the ends under one of the vertical spoke splints. Continue adding splints and snugging them down into place until desired basket height is reached. Clip off spoke splints protruding above last weaver. For the rim use two thicker splints, placing one inside the last weaver and the other encircling it on the outside. Secure these by wrapping the rim with split spruce root, basswood cord, or a very thin ($1/4$ inch) ash splint. Wrap another one in the opposite direction to create a crisscrossed pattern. Get fancy and try making one with a stout handle. You are now ready to gather blueberries in your authentic Ojibwa ash basket.

The Black Ash twig and bud resembles a miniature deer hoof and leg.

Trees of the North Woods and Boundary Waters

Family Pinaceae (Pines)
- ❏ Balsam Fir   *Abies balsamea*
- ❏ Black Spruce   *Picea mariana*
- ❏ Tamarack   *Larix laricina*
- ❏ Jack Pine   *Pinus banksiana*
- ❏ Red Pine   *Pinus resinosa*
- ❏ White Pine   *Pinus strobus*

Family Cupressaceae (Cypress)
- ❏ Northern White Cedar   *Thuja occidentalis*

Family Betulaceae (Birches)
- ❏ Paper Birch   *Betula papyrifera*

Family Salicaceae (Willows)
- ❏ Quaking Aspen   *Populus tremuloides*

Family Rosaceae (Roses)
- ❏ Pin Cherry   *Prunus pensylvanica*
- ❏ Juneberries   *Amelanchier* species
- ❏ American Mountain-Ash   *Sorbus americana*

Family Oleaceae (Olives)
- ❏ Black Ash   *Fraxinus nigra*

Shrubs

Sweet Gale

Myrica gale

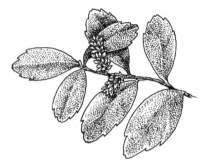

Abundant shoreline shrub 2 to 4 feet tall

Crushed leaves are very fragrant (smells like bayberry)

Bur-like yellow-green fruits

Male and female flowers on separate plants bloom before leaves open

Very early, if not the first, flowering canoe country plant (May 10 to 20)

Other names

　bog myrtle

　bay bush

　sweet willow

　meadow-fern

　golden osier

　pors (in Swedish and Norwegian)

Sweet Gale is a ubiquitous shrub in the canoe country, almost entirely rimming the shoreline of most lakes. Two to 4 feet tall, it is the "ankle biter" that scratches shin and thigh when clambering from canoe to shore. But before cursing, stop, relax, crush a leaf or two, and take a deep breath, inhaling the spicy aroma of Sweet Gale.

Few plants range as far and wide as *Myrica gale.* It is abundant from Newfoundland to Alaska, down to Washington, and south and east to Virginia. Impressive as its North American range is, it doesn't stop there, but circles the northern forests through Europe and Asia.

There are boy Sweet Gales and girl Sweet Gales. Female flowers resemble tiny tufts of red hair at the ends of branches. On separate plants, male flowers, $1/2$-inch long and $1/4$-inch wide, hang conelike from stems. Both flower in mid-May before the leaves open. As with Beaked Hazel, which has a similar floral arrangement, wind is most likely the agent of pollination. This is good, for most insects have not yet emerged, hatched, or thawed from their winter dormancy by early to mid-May. In fact, Sweet Gale is one of the earliest blooming plants in the North Woods.

By late summer, male flower buds have already begun forming. Don't confuse them with the pale yellow-green bur-like fruits or nutlets. Also note the golden resin dots speckling the upper surface of the leaf.

"Fixing" a flat soil

Sweet Gale is a nitrogen fixer like the legumes (peas, beans, and clover), replenishing nitrogen in the soil by trapping it from the air and "fixing" it into organic compounds through certain bacteria contained in nodules on the root system. But unlike the legumes, Sweet Gale does not use *Rhizobium* bacteria; rather it is in partnership with actinomycetes (moldlike bacteria). It accomplishes the same thing, however, creating soil-nitrogen from air-nitrogen so plants can use it for growth. In return, the bacteria gets free room and board, living happily in the root nodules that it helped to create. Because of this unique ability, Sweet Gale has been planted on the nitrogen-poor peatlands of the west coast of Scotland. In America, farmers use industrially fixed nitrogen mixed with fertilizers to replenish their fields.

Finally! A good smelling delouser

The pungent spicy aroma of the crushed leaves is unforgettable and slightly reminiscent of something you've smelled before—like the last time you were in a boutique that carried "bayberry" candles. In reality, Sweet Gale is used to scent these candles, which are named for their close relative Bayberry (*Myrica cerifera*). The catkins, when boiled, provide a scumlike tallow that is skimmed off to make an aromatic wax for candles.

Sprigs of dried leaves have been used for centuries in Europe to delouse beds, scent linens, flavor roasts, and drive out fleas. And like those of Eastern Red Cedar (*Juniperus virginiana*), the leaves are reputed to discourage clothes moths. The nutlets have been used as a sagelike spice.

 Sparky says: Bring a bayberry candle along on your next North Woods canoe trip. Compare the smell of it with the crushed leaves of Sweet Gale. Take a few sprigs home. Tie several in a bundle and hang them upside down to dry. Place a sprig between stored blankets, wool clothes, or linens to keep moths at "bay" and "spice" up any stale odors.

Crest o' the Campbells

The Campbells, an old Scottish Highland clan, display Sweet Gale on their family crest. And proudly so, I might add.

Beaked Hazel

Corylus cornuta

Very common North Woods shrub

3 to 9 feet tall

Male catkins and tiny female scarlet threadlike flowers on the same bush

Flower in early to mid-May

Paired bristly nut cases formed by early July

Other names

    beaked filbert

    hazel brush

Beaked Hazel is one of the most common North Woods shrubs and probably one of the most cursed and hated plants. Anyone who has ever bushwhacked cross-country through the tangle of boreal underbrush knows what I'm talking about. A mere mortal can only be whipped in the face so many times before losing all composure. Even the deer cover their ears.

Slender and smooth light-colored twigs reach from waist high to well over head height. A member of the birch family, Beaked Hazel has 4-inch-long leaves, pointed and finely toothed, but heart-shaped at the base. The male and female flowers appear before the leaves, in early to mid-May. Inch-long catkins, the male flower, hang down and spread their pollen as the wind blows. Hopefully, a few of the four million grains that each catkin possesses will land on the scarlet threads that form the stigma of the female flower, thereby pollinating it. But why the brilliant crimson color? Usually color serves to attract pollinating insects or birds, but Beaked Hazel is entirely wind pollinated. By early July the flower has ripened to a long-husked pair of hazelnuts. Green and covered in bristles that stick in your fingers when you pick them, the nuts are in high demand by the North's Red Squirrels.

Solid stands of hazel brush form primarily under deciduous stands of Quaking Aspen, Paper Birch, and Red Maple. The more southern American Hazel (*Corylus americana*) has gland-tipped hairs on the twigs and rounder nut husks.

Peanut butter substitute

A nut that is 25 percent protein, 60 percent fat, and the rest carbohydrates is bound to be highly sought after by the wild critters. Red Squirrels waste no time in snatching them up and caching them for winter energy. There are accounts of early Indian groups raiding these caches and replacing the nuts with a handful of corn. Johann Kohl writes about the Wisconsin Ojibwa in 1860: "They also carefully collect the wild hazelnuts, rival the squirrels in their search for them, and keep them in bags. They use them, to some extent, instead of butter, for they often eat them with their bread, or the unsalted maize cakes, to which the powdered nut gives flavor." The Indians used several ways to husk the tough shell. Some buried the hazelnuts in mud for about twelve days until the sheath rotted off.

The familiar sweet filbert is from the closely related tree *Corylus avellana*. Oregon orchards are America's largest producer of such varieties as Barcelona, DuChilly, Alpha, Clackamas, and Daviana. Most of the world's filberts are grown on *C. maxima* in Europe, where yields can reach three thousand pounds per acre.

Paired hazelnuts are encased in long husks.

 Sparky says: This summer, race the squirrels for the hazelnut prize. Pick a peck (or just a handful) and bring them home to dry. Spread in a cool, dry place or hang in a cloth bag for several months. Hammer the sheaths off. Coat the 1/2-inch-high nuts lightly in salt and oil. Now roast on a baking sheet at 350° Fahrenheit for about 1/2 hour, or until a crispy light brown. Roasting on an open fire may be something to try on those days when Jack Frost is nipping at your toes. Hey, who needs chestnuts?

What's that nutty odor you're wearing?

Hazelnut oil is used as a base for many commercial perfumes.

Super sketch

The wood makes a superb drawing charcoal for artists.

Speckled Alder

Alnus incana subsp. *rugosa*

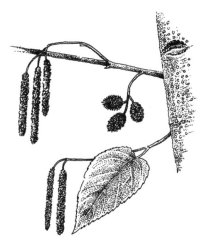

Shrubby tree forms dense tangles in wet areas

Lustrous bronze bark marked with light lenticels

Tiny woody cones and catkins hang off branches

Pollen shed from catkins in early May

Other names

   tag alder

   hoary alder

   rough alder

   gråol ("gray alder" in Norwegian)

   verne (Quebec)

   aulne blanch âtre (Quebec)

Alder swamps can cover extensive areas in the North, forming impenetrable tangles that thwart movement by people but provide a safe haven for certain critters. Moose cows seek out such places for calving in the spring. In the winter, when branches bend low with the weight of snow, Snowshoe Hares move about in the relative safety of the alder maze.

The many-trunked Speckled Alder can reach 20 or more feet in height. Stout trunks are covered with a lustrous bronze bark "speckled" with many horizontal white lenticels. These lenticels are spongy openings that allow the exchange of gases between the atmosphere and the inner plant tissues. Not only decorative but functional!

In late April and early May, before the leaves appear, the male flowers descend in the form of catkins. These tightly packed columns of fragrant flowers shed pollen that pollinates the female flowers, which are found on the same bush and often on the same twig. Fruits ripen into minicones, woody and obvious on the winter plants. Pine Siskins, American Goldfinches, and Common and Hoary Redpolls all feed on the seeds. Leaves unfurl shortly after pollination. Two to $3^1/2$ inches long, they are variably toothed and pointed.

Speckled Alder forms huge colonies by sending up new shoots from underground stems. Or, as I vividly discovered while cutting ski trails this fall, branches can sprout roots wherever they touch the ground.

Nitrogen junkie

Like the legumes (peas, beans, and clover), Speckled Alder "fixes" nitrogen in the soil. Lumpy root nodules harbor bacteria that capture nitrogen from the air spaces in the soil and transform it into a usable form for plants. This nitrogen is later recycled back into the forest ecosystem through shed leaves. But unlike legumes, the bacteria doing the fixing are not a *Rhizobium* species but rather actinomycetal, or moldlike, bacteria (see Sweet Gale). Actinorhizal plants are usually fast-growing pioneer trees and shrubs. It is said that they influence the accumulation of nitrogen in soils considerably more than legumes, which are chiefly disturbed-ground crops. In other words, though we may curse alder at times, we should take a broader perspective and appreciate it for its crucial role in the nutrient economy of the North Woods.

Today is a good day to dye

Ojibwa women used Speckled Alder to dye Porcupine quills a brilliant scarlet. Their ancient recipe called for a handful of inner bark mixed with two handfuls of Bloodroot (*Sanguinaria canadensis*) root, a handful of Wild Plum inner bark, and a single handful of Red Osier Dogwood inner bark to be boiled together. Alone, alder roots provide a brown color, the leaves a yellow-green, and the bark a yellow-brown.

Ojibwa women used Speckled Alder to dye Porcupine quills a brilliant scarlet.

Sparky says: All plants breathe. Green plants take in carbon dioxide and water to make glucose (plant food) and in exchange give off oxygen as a waste product of photosynthesis. They breathe so we can breathe. Survival would be impossible without green plants. This transpiration of molecules takes place through tiny holes, called stomata, on the leaves' undersides. An amazing one hundred thousand are packed into a mere square inch of leaf. Through these microscopic pores the plant also "perspires" excess water. Small plants throw off half their weight in water each day. In the summer, a full-grown maple tree gives off four gallons of water each hour, or an incredible twelve tons each month! To dramatically illustrate this concept, pick a handful of Speckled Alder leaves or any other green leaves and place them in a clear plastic bag. The optimal time to gather the leaves is after dark, when the stomata are closed so as not to lose precious water overnight. Seal the bag, then let it sit in the sun. Watch as water vapor condenses to form visible droplets on the inside of the bag.

Sap spiles

The Ojibwa used hollowed-out alder stems as sap spiles in their spring maple sugar bushes.

Scratch . . . scratch . . . scratch

Indians and the pioneers cured itching by rubbing the inner bark of alder all over their bodies.

Labrador-Tea

Ledum groenlandicum

Low, evergreen shrub

Leathery leaves are fuzzy underneath

Most common in sphagnum moss beds in Black Spruce bogs and forests

Blooms from late May through the third week in June

Other names

Hudson's Bay tea

muckigobug ("swamp leaf" in Ojibwa)

muskeg tea

Hairy armpits and drooping arms are two excellent field characteristics of this shrubby heath, according to Duluth naturalist Denny Olson. He's referring to the sagging leathery leaves with rolled-under margins and their rust-colored woolly undersides. Leaves are 1 to 2 inches long and borne on woody perennial branching stems 1 to 3 feet tall. Flowers the size of a dime, with five white petals and five long stamens, form a dense terminal cluster, or raceme, with the outer perimeter of buds opening first. Three different shrubs I looked at had twenty-four, twenty-seven, and thirty-one flower buds per head. Open flower buds may occur as early as late May, but June is prime time for this sweet-smelling heath.

Ranging from the Labrador Peninsula (where it was first recorded) across Canada and just tiptoeing across the northern tier of the United States, Labrador-Tea is a North Woods specialist. It chooses the cold bogs and the mossy upland sites under Jack Pine and Black Spruce to thrive in.

Gold!

Lowly Labrador-Tea is being used in northern Minnesota and Canada by prospectors searching for gold. I talked with Larry Zanko, a biogeochemist with the University of Minnesota-Duluth's Natural Resources Research Institute, about the process. First of all, why Labrador-Tea? Some species have physical barriers to gold uptake, but Labrador-Tea does not and will, in fact, concentrate gold in its stems. The second criterion is its wide distribution, which makes it possible to use the process on exploratory sites anywhere on the vast gold-bearing Canadian Shield.

So, how do you use Labrador-Tea to find gold? Gather a bunch of stems, strip off all the leaves and bark, and "ash it" to form a concentrated sawdustlike briquette. Run a neutron activation analysis and look for gold concentrations of greater than 2.5 ppb (parts per billion). Why not give it a try? All the big mining companies are doing it.

Mouse repellent

Linnaeus, the Swedish botanist, reported that in his home country the leaves of Labrador-Tea were stored with corn to keep mice away. If I had corn or mice I might try this.

To tea or not to tea

Every dime-store survival book and scout handbook I've ever read recommends Labrador tea as a hearty drink capable of nourishing and renewing the tired outdoor adventurer. But frankly, I don't like it. It's weak and bitter and just plain tastes bad. Former Quetico Provincial Park naturalist Shan Walshe agreed with me, calling it "a very inferior tea." I know, the Ojibwa drank it. But even they sweetened it with maple sugar (and most likely copious amounts at that!). Swamp water would taste good if you added enough sugar or honey! Oh yes, and I know it was used during the Revolutionary War, but they didn't have anything else; they dumped all the good stuff in the harbor.

I can recommend it for certain situations. Take, for example, the plight of the great eighteenth-century cartographer and explorer David Thompson, who, while mapping the remote Athabasca region of Canada, was forced to eat "the inside fat of slain eagles." He and his partner became very sick. Writing in his journal, *Travels in Western North America: 1784–1812,* he notes that "in the night we were both awakened by a violent dysentery [i.e., explosive diarrhea] . . . I filled the pewter basin with Labrador-Tea, and by means of hot stones made a strong infusion [i.e., tea], and drank it as hot as I could which very much relieved me." So the next time you get sick from eating the fat of dead eagles in the hinterlands of Canada, give it a try!

But seriously, drinking too much Labrador-Tea can cause headaches, cramps, paralysis, and intestinal problems due to small amounts of andromedotoxin, a poisonous compound found in the leaves.

 Sparky says: Pluck a good fistful of the "hairy armpit" leaves of *Ledum groenlandicum* and decide for yourself as to the qualities of this North Woods tea. Steep fresh or dried leaves in boiling hot water for 5 minutes (no longer). Sample and then sweeten to taste. Use restraint and try to limit yourself to just one cup. I know this might be tough, but remember what I told you about those nasty andromedotoxins.

Leatherleaf is the most abundant bog shrub in the Boundary Waters and Quetico, yet it is rarely admired or, for that matter, even recognized. "Seas of Leatherleaf" cover openings in bogs. But alas, it is only cursed, or worse, ignored. When it sports beautiful nodding white bells early in the canoe season, few are traveling.

Profusely branched shrubs reach 2 to 4 feet. Evergreen, paddle-shaped leaves are alternate and get smaller as they climb the stem. Scurfy brown undersides are visible from a distance. Flowers hang down in a row from the axils of the tiniest stem-top leaves. The stem usually ends up horizontal, so the dozen or so nodding white bells with five fused sepals and ten stamens look as if they're pinned to a Lilliputian clothesline. One of the earliest North Woods bloomers, the flower buds (formed the summer before) open in mid-May when their roots may still be encased in ice. Flowers last into early June.

A wet desert

The heath family (Ericaceae) includes many North Woods species that are adapted to survive in acid soils and the extremely acidic conditions of the bog. Included in this group are Leatherleaf, Bog Rosemary (*Andromeda glaucophylla*), Bog Laurel (*Kalmia polifolia*), Labrador-Tea, Creeping Snowberry, Small-fruited Bog Cranberry, Wintergreen, Bearberry, and Lowbush Blueberry. Bogs are wet deserts, containing water so acidic that roots don't readily absorb nutrients. Survival techniques of heaths include evergreen leaves, which save energy by not having to grow all new ones every spring, and tough, leathery leaves with waxy surfaces or woolly undersides and sunken stomata, which conserve water during the cold drying winters and hot summer days.

Bog heaths also have a very specific annual flowering order (bloom phenology). This, in the words of author Janine Benyus, ensures that pollen is "delivered to the right address." If all species bloomed simultaneously, pollinating insects would not be restricted to one plant species, and fertilization rates would

Leatherleaf
Chamaedaphne calyculata

Abundant shrub of bogs and lakeshores, often forming huge mats

Paddle-shaped evergreen leaves have rusty undersides

Tiny white bell-shaped flowers hang down from stem tips as if pinned to a clothesline

Early bloomer: May 10 to June 14

Other names

cassandra

be much lower. As it is, average bloom dates are five to fifteen days apart:

Leatherleaf .... 5/10 to 6/1

Bog Rosemary .... 5/20 to 6/10

Bog Laurel .... 5/25 to 6/15

Labrador-Tea .... 6/1 to 6/25

Small-fruited Bog Cranberry .... 6/15 to 7/5

 Sparky says: Give the art of drying flowers a shot. Each species has a proper time to be harvested. Cattails (*Typha* species) should be plucked in early July so the heads remain solid and won't shed their down. Pick goldenrod just prior to peak bloom. Jim Gilbert, naturalist at the Minnesota Landscape Arboretum, also suggests garden flowers such as Brown-eyed Susans (pluck petals off and dry centers), hydrangeas, strawflowers, zinnias, daisies, and baby's breath. Experiment with grasses, berry clusters, club mosses, and seedpods of water-lilies and Wild Irises. Pearly Everlasting (*Anaphalis margaritacea*), as its name suggests, dries well, remaining white and papery just like it was when alive. Tie plants in small bunches and hang upside down in a dim and dry place. Create your own floral arrangements and wreaths. Experiment. Have fun!

Hiking up the parched side of Dutchman Peak one August evening, I noticed the radiant red berries of the Bearberry. Newly ripened, they stood out from the background of withered aspen leaves and spent juneberry bushes. It had been a dry summer. I picked one, as I always do. They look so irresistibly inviting, plump and red, but the taste hasn't changed. They're still mealy and tasteless just like the last time I couldn't resist. Ruffed Grouse like them, though, as dessert after a main course of Quaking Aspen flower buds. The berries survive through the winter, a fact deer take advantage of, if they can find them under the snow. Oh yes, bears also eat Bearberry berries.

The Bearberry belongs to the Ericaceae, or heath family, along with the Lowbush Blueberry, Labrador-Tea, Wintergreen, and Trailing Arbutus (*Epigaea repens*). A low sprawling shrubby plant with shoots growing to only 6 inches high, Bearberry is a true ground hugger that forms mats on rocky ground with reindeer lichen. The small beavertail-paddle-shaped leaves are thick, glossy, and green. The leaves' waxy coating reduces water loss and allows Bearberry to be "evergreen," even in the winter months. The raspberry-size red fruit, which appears too big for such a diminutive plant, can remain "ever-red," staying on the stem year round. It is truly an odd sight to see the bright green Bearberry and its red fruits through the snowmelt of April.

Search for the tiny clusters of pinkish urn-shaped flowers hiding amidst the foliage during the last two weeks of May.

Uva-ursi uses

Bearberry's most famous use by Native Americans was for tobacco. Also known as "kinnikinnick," the dried leaves were crumbled and mixed with the inner bark of dogwood to be smoked in their catlinite (pipestone) pipes. If you've ever tried this blend, you'll know why native peoples made imported Brazilian tobacco one of the top trade items during the North American fur trade era.

Bearberry
Arctostaphylos uva-ursi

Rock-hugging minishrub with reddish bark

Shiny, leathery, alternate, evergreen, paddle-shaped leaves

Cluster of tiny pinkish flowers appearing in mid to late May

Red berries ripen around the first of August

Other names

    kinnikinnick (Algonquin)

    bear's grape

    hog cranberry

    rock berry

    upland cranberry

    mealberry

    sagakominagunj ("berry with spikes" in Ojibwa)

    mjolon (Swedish)

    melbær (Norwegian)

    raisin d'ours (Quebec)

The Ojibwa people smoked the root to attract game. They would follow the tracks, stop, and then smoke again to induce the "four-legged" nearer.

The Cheyenne doctored back sprains using a tea of the leaves. Other tribes treated venereal disease with it.

The fruits of fall

Fall fruits such as mountain-ash berries, Wintergreen, and Bearberries differ from their succulent summer counterparts in having a lower pulp nutrient content, a higher seed-to-pulp ratio, a decreased water content, and an increased fiber content—all factors that allow the fruit to stay intact and edible all winter long, providing a valuable source of winter food to birds and some mammals.

Native Americans smoked kinnikinnick in their pipestone pipes.

Sparky says: Everyone brings their binoculars, camera, and field guides on hikes, but let's put together a "naturalist's discovery kit" that can slip into your daypack and be ready for all your forays afield. Basics should include a small notebook, pen or pencil, 72-inch cloth tape, clear 6-inch ruler with metric and english measures, handlers (or small magnifying glass), pocket knife, several empty film canisters (great for bringing home scat samples!) and a couple of sealing plastic bags. Optional items might include a mini-thermometer, small dip-net, colored pencils, flagging tape, or a flower press.

Organize your discovery tools in a small fanny pack or bag. Compact travel toiletry bags work great. Keep it in the bottom of your daypack and you'll always be ready to make eye-opening discoveries.

"Bear's grape"

The scientific name for Bearberry is *Arctostaphylos uva-ursi*. *Arctostaphylos* is Greek for "bear's grape." *Uva-ursi* is Latin for, guess what, "bear's grape."

No other berry in the North Woods is as highly revered by human or bear as the great wild blueberry. And it does qualify as a true berry. A simple fleshy fruit including a fleshy ovary wall encasing one or more seeds defines the blueberry, along with other true berries, such as tomatoes, bananas, grapes, and gooseberries. You can usually find a ripe blueberry as early as the Fourth of July in the canoe country. But their purple abundance really explodes from late July to early August. The fruit begins as the expanding greenish ovary in mid-June, becoming white, then pink, deepening to blue, and eventually purple. The flowers from which the fruit arises are tiny, urn-shaped, and white, occasionally tinged with pink. They hang in clusters at branch ends and in the leaf axils of new growth. In the hot, dry spring of 1988, Lowbush Blueberry was blooming by May 18 in northern Minnesota. More usual is the last week of May through mid-June. Ants, bees, and blackflies are the agents of pollination.

Lowbush Blueberry plants are many-branched woody shrubs 12 to 20 inches tall. They spread by seed but also via rhizomes, ground-level horizontal stems that sprout genetically identical clones. These colonies can get very large and very old. Electrophoresis has dated one Indian-managed stand in Maine as being 900 years old. But they won't grow on just any old patch of ground. There must be sunlight—lots of it. The soil must be acidic; pH 4 to 5 is ideal. And if recently burned by forest fire, so much the better.

Leaves are alternate, smooth, and minutely toothed, if at all. Velvet-Leaf Blueberry (*Vaccinium myrtilloides*) is very similar but has fuzzy leaves.

The flavor is in the skin

Have you ever wondered why those little pea-sized wild blueberries taste so much sweeter than the fat domestic grocery-store variety? The secret is in the skin. The flavor "cells" are in the skin, and the pulp is just mealy filler. Wild blueberries have a much higher ratio of skin to pulp than domestics and therefore

Lowbush Blueberry
Vaccinium angustifolium

Low woody shrub of open rocky areas, especially old burns

Shiny, alternate leaves are mostly untoothed

Tiny, urn-shaped white flowers hang down in loose clusters

Blooms early, from late May to mid-June

Fat blue-purple berries ripen in early to mid-July

Other names

low sweet blueberry

early sweet blueberry

sweet lowbush blueberry

late low blueberry

whortleberry

bellois (French Canadian)

minagawunj (Ojibwa)

blåbær (Norwegian)

Blueberry flowers appear by late May.

Finding the perfect patch

Have a hankerin' for some wild blueberries but just don't know where to turn? Well, in northern Minnesota the U.S. Forest Service publishes and distributes free maps to picking sites on their lands. The Department of Natural Resources provides this service also. You could ask a local too, but protective pickers are known for their blatant prevarication.

are sweeter overall. It would take 400 wild blueberries to fill a measuring cup, but only 35 to 70 domestic berries.

Commercial growers have been trying to combine the sweetness of small wild berries with the manageability and predictability of domestic varieties for years. Maine, Michigan, Wisconsin, and the eastern provinces of Canada are the main producers. American Indians pioneered blueberry management by using the regenerative power of fire to encourage growth. Some of these original sites are still being used today. Fire rids colonies of old wood, disease, and insects, encouraging re-growth from the intact underground rhizome system. The new stems also produce more blueberries. Come harvest time, some growers use ethylene to encourage abcission, or fruit loosening. Pickers run toothed "blueberry rakes" through the plants, stripping them of their purple bounty. Leaves and such debris are removed later.

Minnesota, on the western edge of blueberry range, is just too dry for commercial ventures. The bushes need at least 1 inch of rain per week. Also, plants produce better after two to four years, at which time the stems are too woody for harvest by rake. A Minnesota hybrid variety, though, called "half-high"— a cross between Lowbush and Highbush Blueberries, is ideal for "pick-your-own" farms, yielding upwards of five pounds of berry per plant.

Crazy cure and cold-weather condiment

Nanaboujou, the Ojibwa god/jokester, personally divulged the cure for "craziness" to his people. Dried flowers of the Lowbush Blueberry were to be placed on very hot stones and the fumes inhaled. Of course, you'd be crazy not to eat the fresh berries too—which the Ojibwa did. They also dried them slowly over a low fire. In winter, when the maple sugar ran low, the dried blueberries became a staple and were mixed in breads and boiled with fish and flesh. Johann Kohl in 1860 wrote of the Lake Superior Ojibwa: "If they wish to give a guest something nice, they take a couple of handfuls of shredded meat, throw it in a saucepan with dried plums or whortleberries [blueberries], and thus produce a soup which restores the strength of a poor fatigued voyageur as if by magic."

Black and blue

Research by the Ontario Ministry of Natural Resources has determined that the blackfly, that cursed and hated blood-licking bug, is the main pollinator of blueberry plants in the far north.

 Sparky says: Canadian blueberry pudding makes a tasty dessert. Enjoy!

2 cups flour
4 teaspoons baking powder
1/2 teaspoon salt
2 tablespoons shortening
1 cup milk
1 cup blueberries

Mix together flour, baking powder, and salt. Work the shortening in with your fingers. Combine milk and berries and mix into dough. Put the dough in a covered metal container and steam for 1 1/2 hours. Dump out and slice. Cover with a topping made of 1/3 cup butter, 1 cup sugar, and 1/2 cup mashed blueberries.

Prickly Wild Rose

Rosa acicularis

Waist-high shrub with prickly stems

Showy pink flower with five petals and many stamens

Wonderfully fragrant

Woods, openings, and rocky ledges

Blooms in June with ripe hips by August

A rose is a rose is a rose is a rose, Gertrude Stein once flippantly remarked. Though there is some truth to that, I wonder if she'd stick by that comment if she held a Prickly Wild Rose in one hand and a cultivated double yellow "gold glow" in the other? Have you ever tried counting the petals on a cultivated rose? You could play "she loves me, she loves me not" for eighteen straight hours! At last count, 421 varieties of rose had been propagated for commercial purposes. Ninety percent of these were domesticated from wild roses of foreign origin.

North America is blessed with twenty-six species of wild rose, which, though more subdued than their cultivated cousins, are nonetheless endowed with a subtle grace and fragrance all their own. (In fact, have you noticed that once something is domesticated, it ends up blockier, bigger, and gaudier than the masterpiece of the original? I have.) Three species inhabit the canoe country. Prickly Wild Rose is by far the most common, but in the wet areas along lakes and rivers grows a smooth-stemmed species known appropriately as Smooth Wild Rose (*Rosa blanda*). Diagnostic of all members of the rose family (which also includes the strawberries, raspberries, cherries, hawthorns, juneberries, apples, and mountain-ash) are the five-petaled white or pink flowers laden with numerous stamens. Prickly Wild Rose petals are large and pink and floppy. Bushes are waist high and covered with prickles, not thorns. (Roses never have thorns. Thorns are actually modified branches arising in the axils of leaves. Hawthorns [genus *Crataegus*] have thorns. Prickles are slender, sharp outgrowths of the bark. Raspberry canes have prickles.

Warm springs can coax Prickly Wild Rose to bloom as early as late May in the canoe country. Late springs may retard blooming until mid-June. Less known for fall color, the compound leaves take on a beautiful red hue in September. In October, when the woods are essentially bare and brown, the rose's remaining soft yellow leaflets on a red petiole stand out. The fruit, called a hip, ripens to a deep red-orange by August but may remain on the bush till the following spring.

Vitamin C in surplus

Rose hips are a red-orange fleshy receptacle which surround numerous achenes (encased dry seeds). There is very little flesh in proportion to seed. Nibbling the mealy fruit down to the seeds is the most expedient way of eating them. But you don't eat them for taste; rather, you tolerate them because they're good for you. Very good for you. Rose hips are one of the best natural sources of vitamin C in the world—ten to one hundred times higher in vitamin C per weight than other fruits. A study on an Alberta, Canada, population of wild roses found that a mere three hips had as much vitamin C as one orange, and a cupful would equal that of a dozen oranges! Hips also contain significant amounts of vitamins A, B, E (seeds), and K, calcium, phosphorus, iron, nicotinamide, organic acids, tannin, and pectin. This veritable pharmacy in red flesh has not gone unnoticed. Rose hip syrup for infants has been popular for decades. During World War II, when citrus supplies were cut off by German blockades, the citizens of Scandinavia and Great Britain were saved from malnutrition and scurvy by the lowly rose hip. Britain's "Operation Rose" gathered and distributed hundreds of tons of rose hip pills and syrup to those in need. Because of this effort, the Wild Dog Rose of the English countryside was honored by being inducted into the Ancient Order of Herbs of Grace.

Hips are best picked after the first hard frost. Snip off the tip and top. Cut lengthwise and scoop out the achenes (seeds). Make a healthful tea by crushing a handful of whole hips, boiling, and straining. Jelly, syrup, and soup are other culinary possibilities.

Due to their low fat content, most birds shun the hips until all their preferred foods are gone. Ruffed Grouse and American Robins, though, are active foragers.

Cataract cure

Ojibwa medicine prescribed a decoction of inner root of wild

Diplolepsis wasp gall on a wild rose stem.

Gall darn

Fifty species of insects make galls on *Rosa* plants. Probably the most common in the North Woods is the globby, peanut-sized stem gall of diplolepsis wasps. In summer this gall is covered by red prickles. In winter look for the numerous exit holes penetrating its smooth woody exterior.

"Itchy-bums"

Children of Canada's northern villages call rose hips "itchy-bums." This describes the symptom that results from an overdose of whole hips, due to the sliverlike hairs of the seeds.

rose and Wild Raspberry to be dripped into the eye to cure cataracts.

Petal power

To this day, native fishermen of Canada's Great Slave Lake chew the petals to a pulp and place on bee stings.

Sparky says: The Swedes also know the pleasures of the rose hip. Try this traditional autumn soup, called *Nypon soppa*. Gather (after the first frost) and deseed 3 cups of rose hips. Boil until tender and squeeze through cheesecloth. Add enough water to bring to 1½ quarts. Place back on the stove and dissolve ½ cup honey and 1½ tablespoons whole wheat flour. Stir and bring to a boil. To be a real Swede, serve chilled, with a dab of whipping cream or ice cream on top and rusks or almond cookies on the side.

The bark of Red Osier Dogwood becomes deeper red in fall and winter. But it is not the only red-barked and oppositely branched shrub in the North. Young Mountain Maple can look very similar. Older and larger Red Osier stems age to pale green and eventually gray in their declining years. The bark also fades once the spring foliage appears. Waist to head height, this many-stemmed shrub can grow in shade or sun but it does like to have its feet just a little bit wet. As indicated by its species name, *stolonifera*, Red Osier spreads by shoots arising from stolons or underground horizontal stems. A trunk in contact with the ground may also sprout roots. This process is called layering.

Oval pointed leaves are borne on the stem in pairs, one opposite the other. They are 2 to 5 inches long. Leaf venation is classic dogwood, with the side veins arching away from the central one and then curving up to follow the margin of the leaf. A rounded to flat-topped panicle of many tiny white flowers faces skyward at twig ends. Each flower has four petals. The annual bloom happens in late May to late June, depending on how much "spring" there was in spring. By late July or early August the berrylike fruits have ripened to white.

Animal allies

Dogwoods need pollinators to bypass self-sterility and dispersers to remove the pulpy flesh, which can hinder germination. Blossoms attract bees, beetles, butterflies, and flies. Cerambycid, or long-horned, beetles are notoriously fond of *Cornus* but actually transfer little pollen. Andrenid bees, halictid bees (sweat bees), and syrphid flies are the chief agents of pollination. The fruit, which is unpalatable to humans, is relished by Black Bears and birds. During migration, flycatchers, vireos, and warblers gorge on the berries. Pheasants, grouse, grackles, grosbeaks, thrushes, thrashers, towhees, bluebirds, juncos, waxwings, sparrows, starlings, woodpeckers, crows, and jays all feed on the fruit. Even Wood Ducks and gulls have been known to sample the dogwood fare. But no matter the animal,

Red Osier Dogwood
Cornus stolonifera

Shrub 3 to 7 feet tall

Stems bright red

Opposite branching

Flat cluster of tiny white flowers blooms late May to late June

White berrylike fruits form in late July to late August

Other names

miskwabimic (Ojibwa: "MIS-kwa-BI-mic")

Dogwood gathering basket

Dagwood?

Do dogs dig dogwood? Or why is it called dogwood? The straight stems of dogwood were used in centuries past to make arrows and daggers. "Daggerwood" may have been corrupted to "dagwood" which was then converted through miscommunication to "dogwood." Well, do you have a better explanation?

none can digest the hard pit in the fruit's center, which is passed in the droppings. This journey through the digestive tract is essential to germination, though, as it removes the pulpy exterior and somewhat scarifies the seed.

Good smoke

The Lake Superior Ojibwa manufactured a fairly good smoke from the inner bark of Red Osier Dogwood. According to Frances Densmore, a cultural anthropologist of the 1920s, they would scrape the inner layer of bark onto a frame, which was then held over a fire to toast. Two parts Red Osier was mixed with a secret ingredient (European twist tobacco) to complete the formula. I have read sources, though, that attribute narcotic effects and stupefaction to Red Osier tobacco smoked in excess.

Sparky says: Try your hand at basket weaving! Gather the longest, reddest, straightest, most uniform Red Osier Dogwood twigs you can find. Bundle, and let them sit for several months. The mistake most people make is trying to weave fresh twigs. The basket may be tight when you finish but will loosen up and fall apart as it dries. When you are ready to start, soak the dried stems overnight in a pond, lake, or bathtub until they regain their flexibility. Now, check out *The Maine Farm* by Joseph Stanley and follow the instructions. Design your own mushroom-gathering basket or fishing creel.

It is only August in the canoe country, but the first signs of fall are appearing; Bracken Ferns are drying brown, Spreading Dogbane (*Apocynum androsaemifolium*) leaves are yellowing, the birds have quit singing, and the Mountain Maple leaves are tinged with orange. It just seems too early, but this is a sign that winter is surely around the next point. It is not too soon to put up more firewood and caulk the cabin's cracks.

The Mountain Maple is a true maple, though it is rarely more than a many-trunked bush growing in the understory. A monstrous Mountain Maple would be 30 feet high with an 8-inch diameter trunk. It is by far the most common maple in the Boundary Waters Canoe Area Wilderness/Quetico lands, and is often the dominant shrub in rich soils beneath Black Ash, Balsam Fir, Northern White Cedar, and White Spruce. The big stands of Sugar Maple that the Ojibwa depended on for sap to make sugar were to the east of the canoe country along the high ridges overlooking Kitchi Gammi (Lake Superior).

Like all its maple cousins, the Mountain Maple has opposite branching, but, unlike theirs, its leaves have only three distinct lobes.

The upright candlelike spikes of fragrant tiny white flowers appear after the leaves are fully formed in early June.

Maple of the Moose

Mountain Maple is also appropriately known as moose maple since these horse-sized herbivores make it a major part of their early summer diet. A Moose-browsed Mountain Maple is a ravaged Mountain Maple.

Sparky says: Keys are an essential tool to hard-core botanists. No, not keys to the lab, or keys to the herbarium, but rather "keys to trees." These keys are written, and successful passage through them leads to correct identifications. Such keys are especially useful when separating several closely related species

Mountain Maple
Acer spicatum

Large, many-trunked shrub

Leaves are three-lobed, turning orange in late summer—early fall

Young twigs are reddish

Upright spikes of fragrant tiny flowers open in early June

Other names

   moose maple

   aninatig (Ojibwa: "A-ni-na-TIG")

   water maple

   low maple

within a genus. I've constructed a simple one here as an introduction to their use. This is a binomial key. If the first "A" is not true for your tree then move on to the second "A." Continue on through the letters, following the true statements, until you arrive at a species name. This should be the identity of your mystery tree.

A-Leaves in needle form

    B-Needles evergreen .... Pines, Spruces, Firs, Cedars

    B-Needles all dropped in fall .... Tamaracks

A-Leaves not in needle form

    C-Branching is alternate

      D-Bark is papery and peeling

        E-Scraped bark of young twigs
          smells like wintergreen .... Yellow Birch

        E-Scraped bark not as above .... Paper Birch

      D-Bark is smooth and not peeling

        F-Leaf petioles flattened and
          longer than leaf .... Quaking Aspen

        F-Leaf petioles not as above

          G-Spring buds resinous
            and aromatic .... Balsam Poplar

          G-Spring buds not as above

            H-Leaf deeply toothed .... Big-tooth Aspen

    C-Branching is opposite

      I-Twigs thick with dark brown buds .... Black Ash

      I-Twigs slender

        J-Grows as single-trunked tree .... Red Maple

        J-Grows as many-trunked
          bush .... Mountain Maple

Family Myricaceae (Sweet Gales)
- ❏ Sweet Gale   *Myrica gale*

Family Betulaceae (Birches)
- ❏ Beaked Hazel   *Corylus cornuta*
- ❏ Speckled Alder   *Alnus incana* subsp. *rugosa*

Family Ericaceae (Heaths)
- ❏ Labrador-Tea   *Ledum groenlandicum*
- ❏ Leatherleaf   *Chamaedaphne calyculata*
- ❏ Bearberry   *Arctostaphylos uva-ursi*
- ❏ Lowbush Blueberry   *Vaccinium angustifolium*

Family Rosaceae (Roses)
- ❏ Prickly Wild Rose   *Rosa acicularis*

Family Cornaceae (Dogwoods)
- ❏ Red Osier Dogwood   *Cornus stolonifera*

Family Aceraceae (Maples)
- ❏ Mountain Maple   *Acer spicatum*

Shrubs of the North Woods and Boundary Waters

Wildflowers

Bullhead Water-Lily

Nuphar luteum subsp. *variegatum*

Grows in shallow still water

Yellow, round flower heads

Large, floating, hoof-shaped leaves

Blooms from early June until early July

Other names

　bullhead lily

　yellow water-lily

　greater yellow water-lily

　large pond-lily

Bullhead Water-Lilies can be found in the calm-water company of Fragrant White Water-Lilies, Tuberous White Water-Lilies (*Nymphaea tuberosa*), and Water Shield (*Brasenia schreberi*). They all need the silty bottom and clear water found along the margins of some North Woods lakes and ponds. Under-muck rootstalks may reach 3 feet in length and are covered with circular scars. Tough but flexible leaf and flower stems arise from these rootstalks. Resting atop the water's surface are the moose track-shaped leaves, cleft only one-third of the way through. Bullhead Water-Lilies would rather be a little closer to shore than their white relatives, preferring 3 feet or less of water.

The yellow, globular flower heads give one the impression of a flower that has never fully opened. But that's all the further they spread—2 inches across, 3 inches max. The cup-forming petals are actually fleshy yellow sepals concealing the true stamenlike petals within. An umbrella-shaped stigma dominates the center of the plant. I have seen Bullhead Water-Lilies blooming as early as June 8 and as late as July 9.

Two other subspecies of *Nuphar luteum* are known from the canoe country. A smaller-leafed (4 inches across), smaller-blossomed (1 inch across), bullhead-lily with a red disc (umbrella portion of the stigma) is known as Small or Least Yellow Water-Lily (*N. luteum* subsp. *pumilum*). And wherever this subspecies and Bullhead Water-Lily cross paths there is always the possibility of producing a hybrid known as *N. luteum* subsp. *pumilum*x subsp. *variegatum*. It retains the large size of *variegatum* but possesses the red disc of *pumilum*. Common names used to describe this hybrid are Lesser or Red-disked Yellow Water-Lily.

Cross-pollination insurance

When the Bullhead Water-Lily flower first begins to unfurl, the only opening is a tiny triangular orifice directly over the stigma. As a specific species of bee squeezes through the hole, it is forced past the stigma. The next day the flower opens, revealing the anthers, exposing the pollen and ensuring cross-fertilization.

Of humans and deer

The starch-packed rootstalks are relished by White-tailed Deer. And those rodents with foresight, Muskrats and Beavers, cache them in their homes for winter snacking. In fact, historically, Native Americans commandeered the rootstalks from the Muskrat's house for their own winter use. They boiled the rootstalks and ate them like potatoes. But being high in alkaloids, the roots are very bitter and are not recommended for the palate of modern humans.

Home, sweet home

My friend and colleague, Jari Kouki of the University of Helsinki, Finland, has studied the incredible relationship between bullhead lilies and a species of chrysomelid beetle. The floating leaves of the closely related *Nuphar lutea* provide a home from birth to death for the diminutive Water-Lily Beetle (*Galerucella nymphaeae*). Adults that have overwintered colonize the leaves by mid-May, copulate, and the females lay eggs in batches on the leaf's top side in late May and early June. Twelve is the average number of eggs per batch, and with 10.3 batches per leaf, the average floating leaf is home to 123 developing beetles. What's interesting is that Water-Lily Beetles are herbivores and immediately on hatching go about the business of eating their "home, sweet home." Even though in their combined effort they can consume only 13 to 22 percent of the leaf's upper surface and palisade tissue, the openings they make allow bacteria and fungi to get a foothold. Rapid decomposition ensues, and the leaf sinks. It is now an expedient time for the larval beetles to ABANDON SHIP! Unfortunately, they are created for a terrestrial life, with no adaptions for swimming. So they simply float over to a newly emerged leaf and continue chewing . . . if, that is, they survive the hazardous crossing. Dr. Kouki found that with an average leaf-life of 4.5 weeks and an average beetle development time of 5 to 6 weeks, none of the larva metamorphosed into adult beetles on their

Boils away!

Indian healers used a poultice of roots to treat boils and skin inflammations. A decoction made from dried rootstalk was touted as excellent medicine for asthma, chest pains, and heart disease.

natal leaf. Emigration is a way of life for Water-Lily Beetles, and a very risky one at that.

 Sparky says: Do you have a spare patch of ground in your lawn or on your land that you could donate to wildlife? If "yes" is the answer, then try your hand at water gardening in a hand-dug frog pond. Ideally, choose a spot into which water naturally drains. Dig out a depression that slopes from 3 feet deep to 1 foot deep. Sculpt shelves from the pond walls where potted aquatic plants can sit. Don't limit yourself to a traditional oval shape. Be creative! Size is limited only by the size of your liner. Next, distribute a 3-inch layer of sand over the bottom and sides. Lay on your thick poly liner. (Commercial ones can be very expensive. A friend suggested using old waterbed liners as an alternative. Give it a try!) Now cover the liner with sand and anchor the edges with rock. Fill with a garden hose or let the pond fill naturally. Plant the seeds of native aquatic plants, such as Bullhead Water-Lily, or try some commercially available exotic water-lilies such as lotuses. In early summer they can be potted and set on your underwater shelves. Come fall they can be removed and stored over the winter indoors. There are many catalogs available that specialize in plants for water gardening. Addition of a submersible pump lets you create recycling waterfalls. But the real joy may be in all the critters your pond will attract. Frogs, toads, salamanders, bathing birds, and drinking mammals will all appreciate your efforts.

Loonshit is the favored substrate of *Nymphaea odorata* in the canoe country. You know the stuff: that deep, sucking muck that grabs at your paddle. Actually a rich, organic soil, it is perfect for growth of water-lilies. But they do need fairly clear water above the loonshit so sunlight can reach the developing stem. This factor limits them to lake edges, where the water is 3 to 10 feet deep (occasionally to 15 feet in the clear waters of the Boundary Waters Canoe Area Wilderness and Quetico). Water-lilies cannot tolerate moving water.

The thick rootstalk lies prone beneath the muck, serving as an anchor and sending up long, hollow, flexible flower and leaf stalks to the surface. Leaves are nearly round, with a deep cleft cutting to the center. Undersides are purplish and slimy from adherent algae. Many aquatic critters find a home there too. Floating leaves a foot across are not rare.

Blooming in the North as early as late June and early July, Fragrant White Water-Lilies will continue to bloom into August. Individual flowers, however, flourish for but a moment, blooming three or four mornings in a row and then fading. It is said that the sweet-smelling flowers (less than $4^1/2$ inches across) of *Nymphaea odorata* bloom from 7 A.M. to 1 P.M. and the odorless but larger flowers of the closely related *N. tuberosa* open fully between 8 A.M. and 3 P.M. Late in the day, or on dark, cloudy days, both species close up, guarded around the outside by green sepals. All white water-lilies have numerous lance-shaped white petals encircling a yellow cluster of stamens. Petals are rarely tinged with pink.

The Least or Small White Water-Lily is a pint-sized version of the Fragrant. Leaves are less than 4 inches across and flowers are barely bottle-cap-size. They bloom towards the end of June. Shan Walshe, former Quetico naturalist, claims there is a small-flowered variety of *N. odorata* also (var. *minor*).

After their week or less of glory in the sun, the flower heads of Fragrant White Water-Lilies become much less fragrant and not so white as the ovaries ripen into seed pods. But during this process of senescence, the stem contracts and coils, dragging

Fragrant White Water-Lily
Nymphaea odorata

Large, floating aquatic flower

Found in still water 3 to 10 feet deep

Floating leaves, round and up to a foot across

Flowers 6 inches across, comprising many white petals and a yellow center

Blooms early June to early August

Other names

    white water-lily

    sweet white water-lily

    nokkerose ("rose of Necken," an evil water-spirit, in Norwegian) the fruiting head down into the watery depths. Underwater, the hard-cased fruit develops. Eventually the pod splits, releasing twelve to thirty-five marble-size seeds that float to the surface. Their buoyancy is due to a fleshy attachment called an aril. If they survive foraging ducks, the seeds will eventually become waterlogged and sink into the muck and take root.

Beetle buddy

The active, long-antennaed, ladybug-length donacia beetles have found a very good home on the Fragrant White Water-Lily. Watch for the metallic purplish green beetles as they scatter before your advancing canoe. A long-horned leaf beetle of the family Chrysomelidae, it finds all it needs in all facets of its life on this aquatic plant. Starting life as eggs attached to the underside of one of the large floating leaves, they hatch into white, robust, grublike larva that gnaw on the water-lily's underwater roots, leaves, and petioles. They breathe by inserting their sharp hind-end spurs (actually modified spiracles) into the leaf or flower stems and sucking out air—a kind of a snorkel-type arrangement. During winter, when the plant dies back and is often totally submerged, oxygen levels in the plant tissues decline and *Donacia* species sink into a metabolic diapause, drastically reducing their oxygen requirements. They pupate within an air- and watertight silk cocoon, still obtaining air via plant tissue. Emerging as adults, they make for the surface and can often be found in the depths of the large white flower, feeding on pollen. Search for signs of their presence on, under, and in the Fragrant White Water-Lilies.

The water nymph

The genus name for white water-lilies, *Nymphaea*, is derived from the Greek *nymphe*, a name for nature's minor divinities of ancient mythology. Nymphs were usually beautiful maidens who inhabited the mountains, forests, and waters.

Pink and white emigrants

White water-lilies were introduced to European water gardens from America via England in 1788. One hundred years later, a West Virginia farmer dug up seeds from an old pond area that produced deep pink blooms. These were carefully propagated and today have become one of the most popular garden water-lily varieties in the world.

Uterine cancer cure

The thick, fleshy rootstalk, collected and sliced in the fall, becomes light and spongy in feel but is impossibly bitter to the taste buds. Almost solid starch, the root contains much tannin and gallic acid. Herbalists create curative gargles to treat mouth and throat ulcers and just plain old sore throats. Baldness, the Ojibwa claimed, could be treated using the root. There is also a historical anecdote of a woman being cured of uterine cancer by an Indian woman who administered a root decoction and injection.

Sparky says: Moose make water-lilies a major portion of their summer diet. Deer, Beaver, and Muskrat partake also. Ducks forage on floating seeds. Feeling left out? Then try "pop-lily"! Gather the underwater seed pods in August. Canoe the shallows on a sunny, calm day so you can see the pods. Pry them from the hard case. The seeds, rich in starch, oil, and protein, are now ready to be popped. Fry them in a layer of oil, keeping the pan in constant motion until they are done popping.

Marsh Marigold

Caltha palustris

Grows in standing water of marshes, creeks, swamps, and ditches

Golden yellow buttercup-type flower

Mass blooms are spectacular in the still mostly brown landscape

Very early bloomer: May 10 to May 31

Other names

  cowslip

  capers

  soldier's buttons

  meadow boots

  drunkards

  crazy bet

  bekkeblom (Norwegian)

  kabbleka (Swedish)

Neither a marigold (Compositae family) nor a cowslip (European species of the family Primulaceae), the Marsh Marigold is a buttercup, and like its relatives it possesses a cuplike flower with an abundance of stamens and many pistils (three to twelve). The five rich golden yellow petals are not petals at all but sepals. Flower heads are slightly smaller than a fifty-cent piece.

Enjoying wet feet, Marsh Marigolds push up in the shallow water of swamps, creeksides, marshes, and ditches, often occurring in profuse displays to create a spectacular scene. The overall effect is heightened by the lack of color in the surrounding landscape during the very early bloom period (second week of May to end of May in the North). Leaves are glossy dark green, kidney-shaped, or roundish. The stem is thick, hollow, and succulent.

Groovy UV

What appears to be isn't always so. Such is the case with the yellow flowers of Marsh Marigold. While humans see a solid yellow inflorescence, bees see a flower with a nearly black center and light sepals (kind of like a Brown-eyed Susan). The cause of this is that the flower strongly absorbs ultraviolet (UV) light at the center, and less so on the periphery. Bees and other insects can see UV light, which entomologists believe helps guide them to the pollen and/or nectar. It comes as no surprise then that bees and syrphid flies are the main pollinators of Marsh Marigolds.

Go ahead and kiss a frog

Warts shrink up and disappear if you squeeze a bit of the caustic Marsh Marigold stem juice on them. Verrucaria is the magic substance found in the stems and leaves that is credited with this ability.

Sparky says: On an early canoe trip during fishing opener, cook up some Marsh Marigold greens to go with your fish. Before the plant is done blooming pick the leaves (not the petioles or leaf stems) and boil 20 to 30 minutes in two or three changes of water. Drain, and add butter, salt, and pepper to taste. Serve with Walleye fillets. Caution: Leaves and stems contain an acrid poison when eaten raw. At home try making capers out of the unopened flower buds. Boil 10 minutes in two changes of water and pickle in hot vinegar.

The upper drawing is how we see the Marsh Marigold. The lower drawing shows what bees see when they look at the same flower through UV light-sensitive eyes.

Red Baneberry

Actaea rubra

12 to 24 inches tall

Compound leaves with five leaflets

Terminal cluster of starry white flowers

Clusters of glossy red (or white) berries terminal on stem (late July)

Other names

   doll's eyes

   white beads

   necklaceweed

   trollbær ("troll berry"—a closely related Norwegian species)

Baneberry prefers those preciously rare North Woods open spots along portage paths and hiking trails, taking advantage of the sun to grow to a leafy 2 feet. Leaves are doubly compound and leaflets are pointed, toothed, and deeply cut. A ball-like cluster of "starry" white flowers appears at the top of the flowering stalk in late May or early June. The starry look is the result of flowers possessing stamens that are longer than the petals. Halictid bees (aka sweat bees) perform pollination duties. In late July and early August, the flower ovaries ripen into berries of glossy red or china white, borne on thick red stalks and each tipped with a black spot or "doll's eye." The different berry colors come from two varieties of Red Baneberry, not two species as once thought. Ruffed Grouse, Yellow-bellied Sapsuckers, American Robins, White-footed Mice, and Red-backed Voles allegedly partake of the berries. Many animals have detoxification systems that we lack, allowing them to eat poisonous plant parts and fungi without any harmful effects.

Deadly doll

How could a plant with the innocent nickname of "doll's eyes" possibly be a "bane" to anything? But a very poisonous cardiac glycoside, which is concentrated in the roots and berries (but present throughout the plant), makes this plant anything but a child's toy. In fact, several fatalities have been recorded in children who sampled the bright, candylike berries of a closely related species in Europe. Adults have a stronger defense system, but just a few ingested berries will cause dizziness, colic, and severe nausea. Death is not out of the question either, as people react differently to toxins. But, as with most poisonous berries, their disagreeable taste will discourage casual grazing and likely result in a mouthful being spit, sprayed, and spewn across the forest floor.

 Sparky says: Now that you know to steer clear of the alluring berries of Baneberry, it's time to brush up on other "belly-busters." Make sure you know that the blueberrylike berries of Clintonia are not edible and that Wild Iris leaves and rootstalks contain iridin, a skin irritant and poison. Before downing a quart of Labrador-Tea, realize that too much can be a toxin to your system. Likewise, avoid feeding frenzies on fiddleheads, especially those of the Bracken Fern. Studies show that they can be carcinogenic in large doses. Not to mention the amanita mushrooms, which are responsible for 90 percent of all deaths caused by fungi in America.

Fortunately, there are relatively few poisonous plants in the North, but to know them will give you that extra boost of confidence when nibbling in the woods.

Wild Columbine

Aquilegia canadensis

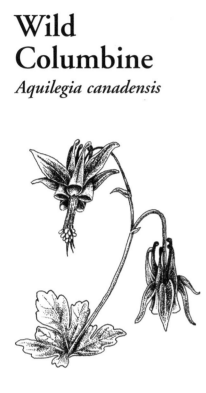

Blooms in June

Showy red and yellow flowers

Prefers shady rock cliffs

Other names

    red columbine

    rock bells

    meetinghouse

    revolver-flower

    lion's herb

    jack-in-trousers

    clucky

    gants de Notre-Dame (Quebec)

June is columbine time in the canoe country, and jaded is the camper who doesn't smile when chancing on a blooming cluster. Scarlet and gold inflorescences hang their humble heads from atop a 2- to 3-foot stalk loaded with compound leaves, divided and subdivided into threes. Cool forest, shaded, and rocky places are preferred. Five pink to deep red petals are modified into upturned spurs, each topped with a tiny bulbous nectar repository. These nectaries are too deep for most bumblebees (*Bombus* species) to exploit, but honeybees (*Apis* species) and halictid bees (*Halictus* species) collect pollen and ensure cross-fertilization. Self-fertilization is unlikely, though, since the numerous yellow stamens (thirty-eight and forty-one in my pressed specimens) descend first and shed their pollen load before the female pistils mature and elongate. Moths and butterflies attempt the awkward upside-down maneuver required to reach the sweet nectar, but Ruby-throated Hummingbirds, with their long retractable tongues, are the master foragers. (Some species cheat and nip a hole through the nectaries, while humans tend to bite off the whole thing!)

Columbine is a perennial plant possessing a stout rhizome. A basal rosette appears in late summer, overwintering in this form.

What's in a name?

"It really does look like a medieval jester's cap!" I thought, after reading such a description. The unique morphology of this flower has inspired a whole handful of vernacular names. "Meeting house" developed because it looks like five heads gathered about a round table; similarly the Latin source *columba*, meaning "dove," refers to the image of five pigeons around a watering hole. Looking into the face of the flower reminded someone of staring into the barrel of an old six-shooter (minus one chamber). "Lion's herb" hearkens back to an old legend that lions garnered strength by grazing a garnish of columbine (I'm not sure where their ranges overlap, but you can't get too picky with legends).

Aquilegia is Latin for "eagle," whose talons reminded the great taxonomist Carolus Linneaus of the spurs of columbine. Interestingly, there was a movement at some point in history to make the columbine America's national flower. A beautiful flower, they reasoned, and with such an appropriate name, incorporating the eagle, our national bird, and columba, as in Columbia, as in the District of Columbia, our nation's capital. (I'm not sure how they dealt with the *canadensis* portion of the Latin name!)

It's not just names that differ from place to place, but also the symbolism and reputation of the plant. Certain North American Indian tribes considered the seeds to be a potent love charm. After dusting his palms with powdered seed, the suitor would hold the hand of the woman he was wooing. And after they washed their hands they lived happily ever after. Contrast this to the reputation of columbine in England, where it was introduced from colonial Virginia during the reign of King Charles I. There it was a symbol of "cuckoldry and a deserted lover," an insult when given to a woman and bad luck to a man.

 Sparky says: Simply sample some sweet spurs this summer. With your teeth, nip off a nectary of a blooming columbine. Leave the leaves alone though, as they contain enough prussic acid to "induce an unpleasant narcotic effect on some." Spread your wild grazing around so as to leave some for the hummingbirds.

Pitcher-Plant

Sarracenia purpurea

Bizarre plant of the bogs

Tubular, vaselike leaves trap insects

Purple, ball-shaped flower heads on separate stalks to 18 inches

In bloom second half of June

Other names

Omukikiwidasun ("frog leggings" in Ojibwa)

Like no plant you've ever seen before, the Pitcher-Plant is a bizarre carnivorous plant of northern bogs. Though we associate Pitcher-Plants with the North, members of the pitcher-plant family (Sarraceniaceae) are best represented in the South. Ten species inhabit the low-nutrient marshes of northern Florida. All members of the family rely on protein from trapped insects to supply their nitrogen and phosphorus needs.

Pitcher-Plant leaves are modified into the shape of a vase, or elongated pitcher, complete with pouring spout. They cluster in a loose rosette around the base of the flowering stalk. Roughly 8 inches tall and 1 to 2 inches in diameter, the "pitchers" are burgandy-veined and green. The less exposed to sun, the less green and more burgundy they become. And like a pitcher they are usually about half full of water.

The single globular flower hardly looks like a blooming flower. All visible parts are burgundy (or wine, or plum, or purple). Note the very large flattened pistil head that effectively shields the stamens. The entire head nods from the top of a separate stalk 12 to 18 inches tall. Look for blooming specimens the last two weeks of June but beware, as the fruiting head looks an awful lot like the flower.

The toothless carnivore

Bogs are highly acidic places. Acidity retards the growth of fungi and bacteria. Fungi and bacteria are the agents of decomposition, which releases nitrogen. Nitrogen is essential for plant growth. Pitcher-Plants grow in bogs and Pitcher-Plants need nitrogen.

Pitcher-Plants' carnivorous nature deals with this dilemma. They simply trap insects in their pitcherlike leaves and dissolve them with enzymes to release the nitrogen locked in their bodies. No problem. Here's the amazing mechanism of the whole process. Insects are first attracted to the pitcher, possibly by its color, but more likely by the odor of decay emanating from it. As they walk down in to investigate, globs of loose platelets stick to their feet, limiting mobility. When they find no rotting meat and turn around to escape, their retreat is blocked by a wall of downward pointing hairs.

Eventually, the insect slips down into the "deadly cocktail," a complex microsystem of rainwater, algae, fungi, bacteria, protozoa, and digestive enzymes secreted by the plant. They drown. Nitrogen is released as the weak enzyme solution dissolves their carcass. (Actually the digestive process is very similar to that of animals.) Special basal leaf cells then absorb the life-giving nitrogen into the Pitcher-Plant's system.

At least that's the conventional wisdom. One researcher did contend that phosphorus was actually the limiting resource in bogs.

The survivors

Incredibly, seventeen species of arthropods survive and thrive in the "death pool" of the pitcher. The Bog Mosquito (*Wyeomyia smithii*) possesses special hooks on its feet enabling it to back down into the pitcher and lay enzyme-resistant eggs. The larvae hatch and frolic and feed amongst the victims. Sarcophagus fly larvae resist digestion by secreting antienzymes. They pupate at the bottom of the pitcher. In some areas the maggot is gathered for fishing bait. Several species of black-and-yellow exyra moths are able to navigate the slippery inner pitcher walls. Hiding by day, the female lays one egg per pitcher. The caterpillars then take over and reroof their new abode. First they cover the pitcher mouth with a dense web of silk. Then they eat the upper inner wall lining so it will die and collapse, further sealing off their home. Now for the doors. The caterpillar cuts one for an exit as an adult moth and another lower down for drainage. It is now ready to pupate in its deluxe cocoon. It emerges in spring as a moth with yellow-veined green wings.

Other species totally avoid the death pool but still use the plant as a hunting stand. Tree frogs occasionally perch on the pitcher's lip and wait for lured insects. And some spiders spin their webs over the mouth.

Newfy flower

Newfoundland has chosen the Pitcher-Plant as its official provincial flower.

Sparky says: Time for a little guided imagery. Lay back and imagine yourself as a fly. You're buzzing about on a nice sunny summer day when you smell something interesting. On closer investigation you determine that the odor is radiating from a cluster of odd tubular leaves. After several overflights you decide to chance a landing. Now on the rim of the tube, you realize that the only way in is to walk down the inner wall. Gently easing your way down you notice that you are losing your usually sure suction-cup traction. Nervously glancing down, you're horrified at the sight of the loose globby plant stuff that's stuck to your feet. As you quickly turn around, it becomes dreadfully apparent that the exit is blocked by down-pointing spikes. Panic sets in, you lose your grip and fall into the stinky death pool below. Drowning, your last thoughts are "Well, at least I'm providing this bog plant with some of its required nitrogen needs." Glug, glug, glug.

Sundews are named for the glistening droplets of nectar ever present on the tips of the leaves and on the petiole's stiff red hairs. But as Duluth naturalist Denny Olson says, "It's not dew . . . it's glue." Insects attracted by the sweet smell or the glistening "dew" are held fast, enveloped, and digested, providing the plant with the essential element nitrogen. Sundews live in nitrogen-poor floating bogs and so must gather it by carnivory.

The basal rosette of leaves is nestled deep in the vegetation of the floating sphagnum mat. Though they're usually near the water's edge, you have to search diligently for them. Most plants' leaves are just boring green "solar collectors," but not so in the sundew. Theirs are also specialized insect traps, and very colorful ones at that. Shaped like miniature wooden spoons, the leaves' petioles (stalks) are quite a bit longer than the roundish fleshy leaf (which is 1 to 3 inches long, with the rounded blade spanning ¼ to ½ inch). Roughly 130 to 260 red tentacles, or glandular hairs, tipped with a secreted droplet of sweet-smelling fluid, cover the entire leaf.

I am describing the flower last because it is much less diagnostic than the basal leaves, which are present for all of the summer. In July the flowering stem grows to pencil height, nodding at the tip. Flowers bloom from the bottom up, one at a time, shyly opening just one petal at a time. A new flower opens nearly every day. The flower in bloom is at the top of the arch, while below it on the stalk are the developing seed capsules, and nodding from the tip are the unopened buds. The five-petaled radially symmetrical flowers are white to pink and may last into early August.

Darwin and the deadly Drosera

Charles "Theory of Evolution" Darwin was fascinated by this ferocious little carnivore with the poetic name. His extensive research culminated in the little known but detailed book *Insectivorous Plants*, first published in 1897. Some of his research is detailed here.

The 130 to 260 fluid-tipped hairs are able to distinguish

Round-leaved Sundew
Drosera rotundifolia

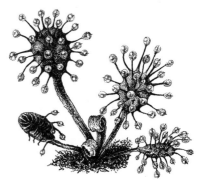

Tiny carnivorous plant of floating bogs

Leaves covered with nectar-tipped red hairs

Blooms in July and early August

Other names

   catch-fly

   rund soldogg ("round sun-dog" in Norwegian)

between "edible" items and "inedible" ones, which do not contain nitrogen. When an appropriate object, say an insect or a spider, lands on the leaf, it is immediately held by the sticky fluid. The tentacles respond by bending inward over the exact spot of contact. Complete closure may take from one to five hours, depending on the age of the leaf, air temperature, and size of the prey. Active victims cause the hairs to close faster. The leaves respond only to sustained touch and not the momentary touch of a raindrop or a neighboring wind-swayed plant. Investigating the ultrasensitivity to pressure, Darwin found that an infinitesimally small piece of human hair, .203 millimeter long and weighing .000822 milligram, was enough to cause inflection of the hairs. This is even more sensitive than the human tongue! When fully closed, the hairs and leaf form a temporary stomach in which digestion takes place.

Darwin was the first scientist to prove true digestion in plants. The acid secreted by the leaf hairs of sundews is very similar to the dissolving gastric juices of animal stomachs. Protein from the body of the victim is broken down into nitrogen-rich compounds that are essential to growth. Darwin even tried placing bits of bone and tooth on leaves, just to see what might happen. The sundew responded by keeping them encased in the temporary "stomach" for six to nine days! On examination, it was found that, indeed, sundew had been able to partially dissolve the bone, soften the enamel, and decalcify both. The lesson to be learned from all this is never take a nap in a bog with Round-leaved Sundew!

Sundew pills

I saw a display of homeopathic tablets at the food co-op the other day. Curious, I picked up the bottle labeled "*Drosera rotundifolia.*" This is what it said: "for temporary relief of dry cough due to minor bronchial irritation which is worse when lying down." I later read in another source that sundews are much higher in vitamin C than citrus fruits. Interesting. I think I'll have to try some "sundew pills" during my next cold.

Autumn hibernacula

Northern sundews form hibernacula (winter buds) in the fall. These are tight clusters of leaf buds that will sprout the basal leaves in spring. In late fall the leaves and roots die back, leaving essentially just the hibernacula, which are able to withstand winter better than an open rosette of leaves.

 Sparky says: Get active! Round-leaved Sundew is a plant that would be very disturbed by disturbance. Fortunately, most of the bogs they grow in are protected from development by their remoteness and inaccessibility. But many other flower species are not so lucky. They have become rare or endangered due to loss of habitat and development. The "prairie garden" has been hit especially hard, but the North Woods is not without its sensitive species. Join an organization dedicated to native plant preservation. An excellent group on a national scale is The Nature Conservancy, which buys threatened ecosystem land outright and, in the process, protects all the unique plants and critters that live there.

Wintergreen

Gaultheria procumbens

Low, woody plant with oval, thick, and glossy evergreen leaves

Large red berry remains on plant through winter

Crushed leaves and berries smell strongly of wintergreen

Grows on dry sites under pines

Nodding urn-shaped flowers bloom late July to mid-August

Other names

   checkerberry

   winisibugons ("dirty leaf" in Ojibwa)

   teaberry

   mountain tea

   ground holly

   deer berry

   box berry

   spice berry

   redberry wintergreen

   petit thé dubois (Quebec) (small tea of the woods)

The heaths (Ericaceae) are a very successful family in the North Woods, dealing well with mineral-poor substrates, acidic conditions, and extreme cold. Creeping Snowberry, Labrador-Tea, Leatherleaf, Bearberry, Lowbush Blueberry, and Small-fruited Bog Cranberry are all members of the Ericaceae. With the exception of the blueberry, all have thick, tough, evergreen leaves to reduce water loss and avoid the nutrient stress of producing new leaves in a very short growing season. Wintergreen's shiny, leathery leaves are oval with a slight point, 1 to 2 inches long, numbering two to five per plant. Crushed leaves smell strongly of, what else, wintergreen. Some leaves fade to crimson in late summer or early fall. Technically a tiny 3- to 5-inch shrub sprouting up along a woody underground stem, Wintergreen enjoys the dry, shaded moss-mats beneath Jack Pines and Red Pines.

In a ecological community where most plants flower in late May or June, Wintergreen is a late bloomer, waiting until late July or even early August to burst forth. But "burst forth" probably is not a proper description for a plant bearing tiny, urn-shaped white flowers that hang humbly from the leaf axils. They are a mere $1/4$ to $1/3$ inch long. By fall they've ripened into a shining red berry $1/4$ inch in diameter. But don't pick them now, as they will plump and mellow over the winter, turning a deeper red in the process. Berries will cling to the plant well into June if not commandeered by hungry Spruce Grouse or chipmunks. "Refreshing" is a fitting adjective for the taste of a spring Wintergreen berry. They make excellent nibbling food.

Wintergreen Certs

The strong taste and aroma of Wintergreen result from the essential oil of wintergreen, the same oil that was used to flavor candies, gum, toothpaste, and cough drops and to camouflage bad-tasting medicines in early America. Oil of wintergreen is composed of 99 percent methyl salicylate and the terpene gaultherilene. Interestingly, Black Birch (*Betula lenta*) and Yellow Birch (*Betula lutea*) contain the exact same fragrant oil. Scratch a young twig end of either species in the spring and your nose will be overloaded with the odor of oil of wintergreen. As one naturalist pointed out, isn't it amazing that two very different plants, one low and creeping and the other a big, deep-rooted tree, can take elements from the vast array available in the ground, water, and air, and produce exactly the same compound, with the only difference being a slight variation in boiling point?

It takes 1 ton of Wintergreen leaves to extract one pound of wintergreen oil! Ironically, modern chemistry has rescued the Black Birch, Yellow Birch, and Wintergreen from overharvesting by the laboratory synthesis of oil of wintergreen. So you'll be happy to know that sucking on a wintergreen Certs is not decimating their populations. Suck on!

Wintergreen and the Boston Tea Party

Have you ever wondered what our forefathers substituted for British tea after the Boston Tea Party? Wintergreen tea, of course! During the boycott of British tea prior to the American Revolution, the colonists switched to this native tea made from the leaves of *Gaultheria procumbens*, Wintergreen. Fortunately, the little heath was quite abundant in southeast Massachusetts (the fruit still being sold in stands of Boston markets in 1924). Prescribed by native Indians and local herbalists for headaches, muscle aches, rheumatism, and colds, wintergreen tea was no secret to them. Today we know salicylates as antipyretics (fever reducers), antirheumatics, and analgesics (pain reducers).

 Sparky says: Experience the smell of natural oil of wintergreen for yourself. If you have Yellow Birch near you, scratch and sniff a twig in spring. If Wintergreen is your neighbor, crush a leaf and let the aroma waft into your nostrils. Bring along something scented with artificial wintergreen. How do they compare?

Creeping Snowberry

Gaultheria hispidula

Creeping leafy stem

Leaves are tiny, waxy, evergreen

White, wintergreen-flavored berries may overwinter

Blooming in last half of May

Other names

   oeufs deperdrix (Quebec)
   (eggs of the partridge)

Creeping Snowberry is a diminutive heath, forming ground-hugging mats in cool mossy pine, spruce, and cedar forests. The creeping stems, edged with tiny (1/4 inch), waxy, football-shaped alternate leaves, flow over moss-covered logs, rocks, or whatever is in its path.

Snowberry's tiny flowers are usually not seen because they are nestled into the moss and rarely searched out. But if you're motivated, look for a white bell-shaped flower with four lobes. They hang singly from leaf axils. A very early bloomer among North Woods flora, Creeping Snowberry flowers during the last half of May.

Like its common name suggests, Snowberry is more recognizable by its berry than by its bloom. Ripening in September, the berries look for all the world like porcelain beads. Lifting the trailing stem out of the clutches of the moss usually reveals a smattering of the relatively large (1/3 inch in diameter) "white-as-snow" berries.

Sparky says: The youngest, freshest leaves make an excellent wintergreen-flavored tea, which is also very high in caffeine. This is crucial information to all you canoe trippers who forgot the coffee on the kitchen counter amidst the frenzy of packing! Quetico naturalist Shan Walshe claimed the berries are a true delicacy when smothered in cream and sugar. (But then again, what wouldn't be excellent smothered in cream and sugar?) Jams and preserves can be made from the berries with the addition of pectin. Good luck, though, in finding enough to make jam.

Cranberries are in the same genus as blueberries, but they are not related at all to the familiar Highbush Cranberry (*Viburnum trilobum*). They grow in cold bogs around the northern hemisphere from Norway to Siberia to Canada. This trailing evergreen with a woody stem up to 4 feet long is dotted with erect stems covered by tiny ($^1/_4$ inch) pointed alternate leaves. Inconspicuous (i.e., crawl through the bog during the height of mosquito season to find them) pink flowers with four swept-back petals grace the top of a flower stalk, which emerges from the tip of the stem. Ripening in late August and early September, the fat $^1/_3$-inch red berry is best gathered after a couple of good frost nips. If left nestled in its mossy home the cranberry will easily survive until the spring, thanks in part to naturally occurring benzoic acid, which is a preservative, and other bactericidal substances.

What would Thanksgiving Day be without it?

Canada had the right idea in designating October the month for the celebration of Thanksgiving Day. Why? Because that is the time of the cranberry harvest, and what would the annual feast be without them? Look on any bag of frozen fresh cranberries, and it will say picked in either Washington, Massachusetts, or Wisconsin. These are the nation's leaders in commercial cranberry harvests. In fact, central Wisconsin news reports have a special cranberry bog forecast to let growers know of impending hard frosts. Ryan Walker, a cranberry farmer from Wisconsin Rapids, said that 1993 was a tough year for growers. The extremely wet summer washed pollen out of the flower, thereby reducing the "set" of the fruit and resulting in a 15 percent drop in harvests. Most of the fresh cranberries in grocery stores were gone by early December, he said. Cranberries are the major fruit crop grown in Wisconsin, with farm gate profits of 75 million dollars, and the finished product bringing 200 million dollars into the Wisconsin economy each year. Only 5 percent of the total harvest is fresh or frozen, while 95 percent is processed into sauces and juice.

Small-fruited Bog Cranberry
Vaccinium oxycoccus

Tiny trailing vine found in floating bogs

Blooms mid-June to early July

Cranberry fruit, up to $^1/_3$ inch across, ripens in early fall

Other names

    crane-berry

    mashkigimin (Ojibwa)

    tranbar ("craneberry" in Swedish)

Crane-berry

The Small-fruited Bog Cranberry was known to early colonists as "crane-berry" in reference to the flower and flowering stalk, which have the crook-necked appearance of the head of a crane. Unfortunately, the "crane" in reference was probably the Great Blue Heron of America or the Gray Heron of Europe, who, unlike cranes, do fold their neck in flight and when hunting. Therefore, *Vaccinium oxycoccus* should be known around the world as heronberry and properly corrupted to "hairyberry." "Hey mom! Pass the turkey, oh, and the hairyberries too!"

The cranberry marsh

It is said that the pilgrims were regaled that first fall by the local Indians with a new and excessively tart berry—the cranberry. It impressed them so much that they chose the tiny red berries over precious Beaver pelts and even more valuable White Pine timber as a sufficiently royal gift for their king in England. It was just a matter of time before people would attempt to commercially harvest and domesticate the wild cranberry, native of the untamed cold bogs of the north. Around 1810 it was first domesticated, and by 1850 the tart berry was gaining a reputation in the New World. German adventurer and ethnologist Johann Georg Kohl wrote in 1858 in his book *Kitchi Gami: Travels with the Lake Superior Ojibway* that cranberries had "recently become a valuable article of export to lower Canada and America, and one of the settlers boasted that he exported several tons annually . . . Half the Indian families then absent from the village [along St. Mary's River] had gone 'dans les ottakas', or to the cranberry harvest." These precious patches were family heirlooms passed down generation to generation like maple sugar bushes. By 1924, fifty million quarts were being commercially harvested annually.

 Sparky says: Making a natural dye can be as easy as boiling up a vegetable for dinner. You'll need an enamel kettle, rinse bucket, colander or cheesecloth, large spoon, measuring cup, stove top, a mordant, and 16 ounces of unbleached wool looped in loose bundles and tied with white cotton string.

First an explanation. A mordant is simply something added to the dye bath to fix the dye on the yarn and make the color more permanent. Old mordants included alder or sumac twigs, blood, urine, lye, and clay. In fact the famous Harris Tweeds were dyed with lichens, using urine as a mordant. A basic recipe is 3 ounces of alum dissolved in $1/2$ cup of warm water or one ounce of cream of tartar in $1/2$ cup of warm water. Iron, tin, and chrome are other modern mordants.

Some North Woods plants good for dyeing and the colors they can produce are listed here: Paper Birch leaves with alum produce yellow. The berries of Staghorn Sumac (*Rhus typhina*) in alum make a warm red. A strong red can be created by fermenting the lichen Rock Tripe in ammonia. Very fresh horsetails with alum produce a yellowish green color, while cranberry juice with alum and tin makes a pinkish tan.

The first step is to chop up your plant parts very fine and place them in a water bath with your mordant. Now dilute the dye bath with four gallons of water. Immerse your yarn and heat to simmer for 1 hour. Let cool in the water. Rinse yarn with cold water until no more color comes out of the yarn. Hang up to dry. Experiment yourself. A good source of info and ideas is Karen Leigh Casselman's book *Craft of the Dyer: Colour from Plants and Lichens of the Northeast.*

Pipsissewa

Chimaphila umbellata

Loose cluster of nodding pink flowers with five waxy petals

Whorls of thick, shiny, evergreen leaves finely toothed around margin

Up to 8 inches tall

Often in dry lichen mats under pines

Blooming last three weeks of July

Other names

    pipsisikweu (Cree)

    gagigebug ("everlasting leaf" in Ojibwa)

    prince's pine

    love-in-winter

    ground holly

    bitter wintergreen

    pine tulip

    ryl (Swedish)

Pipsissewa . . . Pipsissewa . . . The name just rolls off the tongue. This is my favorite flower name in all the North Woods, and it is just a consonant or two away from the original Cree name, pipsisikweu, which literally translates to "it breaks into small pieces." This alludes to the medicinal belief that the leaves had the ability to break down kidney and gallstones. They do not.

I associate the smell of dry pine needles in a sun-dappled summer Jack Pine forest with this flower. It likes dry sites under spruces and pines, occasionally growing amidst reindeer lichen (*Cladina rangiferina*) mats on granite in forest openings. East of Minnesota, it grows under hemlocks (*Tsuga* species). Look for Pipsissewa, a July bloomer, at the height of the short North Woods summer.

Like its close relative One-flowered Wintergreen, Pipsissewa has five waxy petals, nodding flowers, ten prostrate stamens, and a superior ovary. The petals, though, are tinged with rose, stamens are violet, and flowers hang in a loose umbel of two to eight. Thick, glossy, finely toothed leaves appear in one or two whorls about the stem, which arises from a perennial, creeping rhizome. This beauty peaks out at 4 to 8 inches.

Root beer and remedies

Adjectives like "pleasant" and "delicate" pop up in reference to the taste of the leaves, which are touted as a fine nibble for outdoor explorers. In fact, they are a traditional ingredient of root beer. But personally, I found them bitter and quite unpleasant. I'd use adjectives like "icky" and "yucky."

Medicinally speaking, though, Pipsissewa has been used by many cultures for many years. The Mohegans and Penobscots steeped the leaves in warm water and applied them to blisters via compress. The Ojibwa treated sore eyes with eye drops made from a root decoction. The Thompson Indians of British Columbia applied poultices of pulverized plants to reduce swelling in legs and feet. This knowledge was passed from Native Americans to colonists, who continued its use, especially to treat skin irritations and bruises. In fact, Pipsissewa tonic was a staple home remedy in pioneer homes, even making its way into the U.S. Pharmacopoeia from 1820 to 1916. As to the Cree belief that it could break down kidney and gall stones, modern medicine finds this doubtful but does agree that it has value as a mild urinary antiseptic.

 Sparky says: Here's an activity that is both money saving and botanically enlightening. You can't beat that! Next trip to the cabin, camp, or cottage, take a stiff piece of lightweight cardboard and a black pen or pencil. Sit yourself in front of a group of Pipsissewa (or other interesting floral models that catch your eye) and sketch one in detail. Include leaf venation and flower parts. Colored pencils are optional. Jot a note to a friend back home about your day and about your flower. Address it, stamp it, and mail it.

Pyrolas
Pyrola species

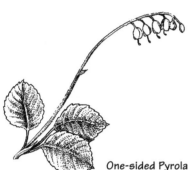

One-sided Pyrola

Basal rosettes of rounded, untoothed leaves

Waxy, cup-shaped, nodding flowers hang off stem

Portages, paths, and campsites in deciduous and boreal forests

Bloom mid-June to late July

Other names

    shinleafs

    vintergronn ("wintergreen" in Norwegian)

Six species of pyrolas inhabit the Boundary Waters/Quetico region. In probable order of abundance, from most to least, they are Nodding Pyrola (*Pyrola chlorantha*), One-sided Pyrola (*P. secunda*), Shinleaf (*P. elliptica*), Pink Shinleaf (*P. asarifolia*), Round-leaved Pyrola (*P. rotundifolia*), and Lesser Pyrola (*P. minor*). Pyrolas are unique in having anthers that do not split lengthwise to release pollen; rather, it comes out salt shaker–style, through the tip. All species described here bloom from mid-June through July.

Nodding Pyrola (aka greenish-flowered pyrola or common pyrola) is arguably the most common boreal pyrola. Found in dry soil under pines, it has relatively large flowers tinged with green. Note also that the petioles are longer than the small basal leaves. Exceptional specimens reach to 12 inches.

Almost as common is One-sided Pyrola, which as its name implies has all its flowering stalks attached to one side of the stem. This often results in the stem bending over from the imbalance. Quetico naturalist Shan Walshe found it "exceptionally abundant" in rich mixed woods of large Quaking Aspen, Balsam Fir, and Paper Birch. It stands 4 to 8 inches tall.

Shinleaf has large elliptical leaves (up to 3 inches) that are flattened on the end and not pointed at all. Their petioles are shorter than the blades. Flower petals seem to be more spread than in other pyrolas. Elongated downswept pistils are also diagnostic. It reaches 5 to 10 inches tall.

Pink Shinleaf is tall and pink—even the stem. Note also the indentations where leaf petiole (stalks) attaches to blade. Flowers are pink and open, and have a long downswept pistil. Not a fan of acid soils, Pink Shinleaf prefers richer clay soils, often at the base of fellow lime-lover Northern White Cedar. It towers over other pyrolas at 8 to 16 inches in height. You'll know it when you see it!

Round-leaved Pyrola is similar to Shinleaf but usually is taller, bears more flowers, and has leaves that are glossier, rounder, and more leathery. Compare leaf petioles too. Those of Round-leaved Pyrola are about equal in length to blades, while Shinleaf's petioles are shorter. It stands 6 to 15 inches tall.

Not yet discovered in Quetico, Lesser Pyrola is also very rare in the Boundary Waters and a proposed "special concern" species in Minnesota. It is a miniature version of Nodding Pyrola, at a diminutive 3 to 7 inches.

Sparky says: Due to the differing bloom phenology and habitat preferences of the pyrola species, It would be impossible to compare them side by side in the field. But take heart! You can do this at your local herbarium. No, an herbarium is not a glorified gerbil cage, but rather a repository for dried and flattened plant specimens arranged systematically. Most universities, forestry stations, natural history museums, and arboretums have one. Call and set up a tour with the curator. They can be great resources for any identification questions you have in the future.

Latin pear

The name *Pyrola* is derived from the Latin for pear, *pyrus*. This refers to the shape of the basal leaves.

One-flowered Wintergreen

Moneses uniflora

Diminutive plant a mere 2 to 5 inches tall

Single nodding flower with five waxy, stiff petals

Basal rosette of evergreen elliptical leaves

Haunts the cool mossy woods, especially under cedar

Flowering period a week either side of the Fourth of July

Other names

    wood nymph

    single delight

    Olavsstake ("Olav's candle-stick" in Norwegian)

    ogonljus ("eyelight" in Swedish)

The personality of this shy little flower seems to be perfectly captured by its common name, wood nymph. Overlooked by many North Woods travelers, it reveals itself to those who can appreciate its fragile beauty and delicate fragrance. Haunting the cool, mossy forests, wood nymph inhabits magical places fit for wood nymphs, forest sprites, and possibly a gnome or two.

One-flowered Wintergreen is a member of the Pyrolaceae family, a group of coniferous woodland flowers with evergreen leaves, nodding inflorescences, and flower parts in multiples of five. The single ivory flowers nod from the tip of the 2- to 5-inch stem, bare but for a minute bract just below the blossom. Ten white stamens surround the green bulbous superior ovary, from which projects a prominent pistil. Five waxy, stiff, and somewhat pointed petals round out the composition of the 3/4-inch head. Flowers are fragrant. The basal rosette is composed of small, lustrous evergreen leaves with a minutely scalloped edge.

Check out the shaded forest floor beneath Northern White Cedars, Balsam Firs, and White Spruce around the Fourth of July for small clusters or single specimens of blooming wood nymph.

Delightful

The genus name, *Moneses*, is derived from two Greek words: *monos* means "single," referring to the single flower, and *hesis* means "delight," which combine to "single delight," a beautifully appropriate title for a delightful little plant.

Sparky says: Seagull Lake's One-flowered Wintergreens first burst forth in 1985 on July 7 and in 1993 on July 2. Other first dates of bloom I've noted in my phenology journal are June 23 and July 2. Most species show a similar narrow window of bloom dates from year to year. Of course, variations in seasonal rainfall and temperature account for minor variations in bloom dates. To keep track of the floral phenology around your home or cabin, note dates of flower blooms, berry ripenings, peak of fall color, sap runs, and pollen shedding on a calendar or in a blank journal. As the years roll by, it becomes increasingly interesting to compare dates. Be diligent and remember, "phloral phenology is phun!"

Indian-Pipe
Monotropa uniflora

Translucent, waxy white flower with no chlorophyll

Grows in deep shade under conifers

Blooms in mid-June to late August

Other names

    corpse plant

    ghost flower

    ghost pipe

    ice plant

    bird's nest

The living corpse of the North Woods, Indian-Pipe is a true oddity in the plant kingdom. This plant gains nutrients from decaying wood and also by parasitizing the root systems of living plants. Because of its method of gathering food, Indian-Pipe has no need of chlorophyll, resulting in its translucent white, molded-out-of-wax appearance. Since photosynthesis is not required, leaves are vestigial, reduced to scalelike growths hugging the stem. The single flower on this 4- to 10-inch plant is 1 inch long and nodding, possessing five petals and ten stamens. After pollination (by whom, is a mystery), the flower head turns upward and the once-succulent plant dries stiff and black.

Why the name Indian-Pipe? Several explanations surface. One relies on the plant's resemblance to an upside-down pipe similar to the English clay pipes traded to the northern tribes for Beaver pelts by the North West Company and Hudson's Bay Company in the eighteenth and nineteenth centuries. The pipe analogy is furthered by the stem's red center, which could be likened to the coal ember in the bowl of a burning pipe.

You may have to push aside some low-hanging leaves or needles to find the "ghost flower," for it truly loves the dark, deeply shaded haunts of the North Woods. It prefers rich, acid soils, so look for Indian-Pipe under conifers, especially in Black Spruce bog forests. In the canoe country, mid-July is the time to look for the first blooming specimens. Indian-Pipe ranges across all of the U.S. and Canada, even dipping south to Mexico. It is also found in Japan and the Himalayas.

Freeloader

Due to its oddball method of securing nourishment, some early botanists deplored this little flower for its "degenerate morals." I suppose this was because it didn't "work" to produce its own food by green chlorophyll and photosynthesis like most "honest" plants. In reality, Indian-Pipe is saprophagous (feeding on dead and decaying organic matter). A true saprophyte depends on wood-rotting mycorrhizal fungi in the soil to transfer nutrients from decaying wood to its own roots. In fact the roots of Indian-Pipe are so entwined in the clump of fungi that they barely even contact the soil. The mycorrhizal fungi, in other words, are the "middlemen" between Indian-Pipe and rotting wood, the "delivery boy" with the groceries, if you will.

Recently it has been determined that Indian-Pipe can also survive and thrive as an epiparasite, living indirectly from the roots of living green plants. In this system the mycorrhizal fungus in the soil acts as a nutrient bridge between Indian-Pipe and the roots of green photosynthesizing plants instead of decaying wood. A *Boletus* species of mushroom (an ectendo-mycorrhizae) has been found to be the "delivery boy" between Indian-Pipe and the roots of pines and spruces. This also helps explain why *Monotropa* species are commonly found in coniferous forests.

Clear sap—clear eyes

The Ojibwa of northern Minnesota believed that by applying the clear sap of Indian-Pipe to the eyes one could clear up cloudy vision.

 Sparky says: Here's a family activity to try on your next North Woods outing. Each person chooses a plant that's special to them. (Don't pick it, just choose it.) Now write a poem in the cinquain form about it. A cinquain starts with a one-word name for the plant. The second line consists of two descriptive terms, the third of three action adjectives, followed on the fourth line by four feelings for the plant. The cinquain ends with a one-word synonym for the plant. Here's an example:

Indian-Pipe
waxy, white
stout, straight, silent
ghostly, haunting, clammy, eerie
corpse-plant

After composing your cinquain, get together and share your creations. For variation, try writing one as a group, or have everyone write about the same species. Note differences in people's perspectives.

Starflower
Trientalis borealis

Early woodland flower of the deep and shady places

Delicate white flower has seven petals, seven sepals, and seven stamens

4- to 9-inch-high plant has a whorl of five to nine lance-shaped leaves

In bloom the end of May through mid-June

Other names

skogstjärna ("wood star" in Swedish—a closely related species)

Starflower is a shy herb of cool and mossy northern forests around the world. In Sweden it is called *skogsjärna* (skoog-shar-nah), which means "wood star," a very lovely and appropriate name. One or two star-shaped white flowers are held upright above a whorl of five to nine lance-shaped leaves. Each blossom is composed of seven sepals, seven petals, and seven stamens, a very rare combination in nature. Starflower blooms from late May to mid-June.

Perennially springing from a creeping rhizome, Starflower is associated with Bunchberry, Clintonia, and False Lily-of-the-Valley. It gets its energy boost for an early bloom by storing nutrients through the winter in fleshy underground tubers.

Holy heptandra!

Seven-parted flowers are very rare in the plant kingdom. But starflower, which has seven each of stamens, petals, and sepals, is such an herb. Swedish taxonomist Carolus Linnaeus, in his original classification of plants by "sexual system," placed Starflower in class Heptandra, a very small group of seven-parted species.

Sparky says: Linnaeus is practically the only botanist I've talked about in this book. So it's high time we all learned about other influential plant people. Pick one and research their life. Where did they do their field studies? What influenced them to get into botany? What contributions did they make? Are there any plants named in their honor? Here's a few names to get you going: Mark Catesby (1682–1749), André Michaux (1746–1803), Pedr Kalm (1715–1779), Thomas Nuttal (1786–1859), and Olga Lakela (1900s).

Wild Strawberry
Fragaria virginiana

This is one of the wild varieties from which the cultivated strawberry was bred. It is very common in clearings and woods, along portage paths, and edging campsites. But it must be searched out, as it often grows tucked down under the surrounding vegetation. And being only 3 to 6 inches tall doesn't help either. Bending down, though, is not considered hardship to those in search of the exquisitely sweet "berries" of June and July, more delicate and fragrant than any fruit should be allowed to be. But what you eat is not a single berry but an enlarged fleshy flower center (receptacle). The "seeds" dimpling the outside are actually the individual one-seeded fruits (achenes). Wild strawberries are but a fraction the size of domestics and yet possess many times the flavor. They nearly melt in your mouth.

The flowers that become the "straw-berries" are typical rose family flowers. They have five green sepals, five white and round petals, and numerous stamens, and are about an inch across. Blooming in the North is a late May and June event. Distinct leaves bearing three coarsely toothed leaflets arise separately on a hairy petiole. They turn a glowing scarlet in late summer. Plants spread by runners (above-ground stems or stolons) and, when conditions are right, are able to colonize areas quickly. One study found that in three years one plant spawned two hundred plants and covered seventy times the original area!

The Woodland Strawberry (*Fragaria vesca*) is also common in the North Woods, preferring dry, bedrock ridges in the Boundary Waters and Quetico. The easiest to see distinguishing characteristic of Woodland Strawberry is the seeds, which are scattered over the nearly smooth fruit surface. In Wild Strawberry fruits, the achenes are embedded in pits on the fruit.

God's berry

"Doubtless God could have made a better berry, but doubtless God never did," remarked the English writer-fisherman (or is that fisherman-writer?) Izaac Walton, sometime in the 1600s. Three hundred years later, I think most would still agree.

Three toothed leaflets on hairy stalk

White flowers with five round petals on separate stalk

3 to 6 inches tall

Flowers late May through June

Berries by mid-June and into July

Other names

common strawberry

meadow strawberry

wild field strawberry

odeiminidjibik ("heart berry root" in Ojibwa)

jordbær ("earth-berry" in Norwegian)

smultron (Swedish)

Cultivation has "added nothing to flavor and much to size." And I agree with a fellow naturalist who proclaimed, "I would rather have one pint of wild strawberries than a gallon of tame ones." The Ojibwa, famous for drying many fruits, only ate strawberries fresh. Whether this was due to characteristics of the fruit that made them difficult to dry or whether they just couldn't keep them out of their mouths, I don't know. Some people with huge amounts of self-control manage to pick and save enough for pie, shortcake, or jam.

The "straw" in strawberry

Where did the "straw" in strawberry come from? Most agree that its origins are from old England. The Anglo-Saxon "streow berie" may refer to the runners "strewn" all over the ground. Or the runners themselves may have resembled straw to an early observer. Straw was used under the plants to protect the berries from dirt and mud. Then there was the old European practice of stringing the fruit on grass or straw. So, take your pick, or make up your own story.

Sparky says: Say "cheeeeese!" So you say you want that Pepsodent smile without the additives and slick packaging. Then try the strawberry, nature's oral cleanser. It cleans, whitens, dissolves tartar, and removes plaque. Rub the juice of fresh berries on your "not-so-pearly-whites" and leave for a few minutes. Rinse with warm water mixed with a bit of baking soda. The other eleven months of the year, when wild fruit is not to be found, substitute Fowler's Extract of Wild Strawberry, available at most co-ops and health food stores.

Wild Raspberry
Rubus strigosus

Arising from a perennial stem as clones, first-year canes, called primocanes, do not flower and have five larger leaflets. Plants mature in the second year (floricanes) and produce white five-petaled flowers with many stamens. Once pollinated by bees (usually), the ovary ripens into an aggregate of drupes we call a raspberry. This fruit, unlike blackberries and dewberries, drops off its receptacle when fully ripe. Very similar to domestic raspberries, Wild Raspberries differ only in smaller fruit size and fewer berries per cane. "Prickly" stems theoretically protect plants from grazing by deer and bunnies, but both animals seem to do a number on winter stems. The stems are, however, pretty good at snagging pant legs.

Logged, burned, graded, and bulldozed sites are all perfectly suited to the Wild Raspberry. Tolerant of a wide range of pH levels (pH 5.5 to 7), all it asks for is a little disturbance and a lot of sun. Open rock faces in the canoe country also sprout raspberry canes.

June is flowering month for the red Wild Raspberry in the North Woods. July and the first half of August is "aggregate of drupe" gathering time.

Berry, berry, good

Eastern and Least Chipmunks, Ruffed Grouse, Blue Jays, Cedar Waxwings, Veeries, Common Grackles, Purple Finches, American Robins, Gray Catbirds, and Rose-breasted Grosbeaks all flock to the juicy ripe fruits, and so do I. But mankind hasn't always shared the wild critters' enthusiasm for raspberries. Evidently, the fruit was essentially ignored in Europe until the early seventeenth century. In America, the Ojibwa boiled up bunches of berries, then let them sun-dry on sheets of birchbark. The little cakes were then stacked and tied in bundles for winter use. But why wait? Doesn't a big bowl of just-off-the-cane raspberries smothered in heavy cream sound good right about now? Monroe Sprowl, in his southern drawl, I'm sure would agree:

Red fruit ripens in July and easily separates from receptacle

White five-petaled flower with many stamens blooms in June

Woody canes 2 to 4 feet tall and covered with bristles

Textured, compound leaves with three to five leaflets

Grows in sunny, disturbed sights

Other names
 red raspberry

If ever I dies an' yo ain't certain I's dead,
just butter some biscuit an' new made bread
An' spread 'em all over with raspberry jam,
Then step mighty softly to whar I am
An wave dem vittles above my head,_
If my mouf don't open, I'm certainly dead.

Leaves and labor (the childbirth kind)

"Pregnant women the world over swear by the leaves of this humble plant. Used by women in China, Europe, and North America, raspberry leaves prepare the womb for birth." So started the raspberry lecture of the "Herbology throughout the Reproductive Cycle" 1993/1994 conference sponsored by the American College of Nurses and Midwives (ACNM). The leaves, harvested before the fruit ripens, contain fragerine, a well-known uterine tonic, which tones the pelvic and uterine muscles. Dr. Violet Russel, in a letter to the distinguished British medical journal, *The Lancet*, said of raspberry-leaf tea: "Somewhat shamefacedly I have encouraged expectant mothers to drink this infusion. In a great many cases labour has been free and easy from muscular spasm." The tea is made by steeping 2 teaspoons of fresh or dried raspberry leaves in 1 cup of boiling water for 10 minutes and then straining. Use the smallest, youngest, greenest leaves and tear into pieces. The ACNM recommends 1 cup per day for the first two trimesters and 2 to 3 cups per day during the last trimester. Consult a physician first.

Sparky says: Try this refreshing red raspberry recipe on your next summer trip to the cabin.
 Mix in a bowl 2 cups of fat raspberries, 1 cup of cream, a few drops of vanilla, $1^{1}/_{2}$ tablespoons sugar, and 1 cup milk. Put all in a blender (or hand-whip at those "primitive" cabins), and serve in glasses with a few raspberries on top.

Thimbleberry
Rubus parviflorus

Huge maple-shaped leaves on a knee- to waist-high plant along the portage are what will probably first alert you to the famed Thimbleberry. As a member of the rose family, it possesses the five-petaled, many-stemmed, white flowers of its kind. Thimbleberry flowers are flimsy but large, each petal spanning nearly an inch. Leaves may be 8 inches across.

The red, thimble-sized fruits are the feature that really impresses me about this plant. They look like a large, deep red raspberry with small drupelets. They are large enough to actually cap the end of any finger. Unlike a thimble, though, they would not be of any use in protecting the skin as they are one of the softest, most delicate fruits in the North Woods. In fact, after filling a small container you'll find that the bottom fruits have been turned to mush. Their taste is wondrous to some (like me!), but just too tangy for others. Unfortunately, it's difficult to find large patches of fruit-bearing plants. Sixty species of birds enjoy the berries of *Rubus* species. Thimbleberries ripen in August.

Gally!

On the stems of some Thimbleberry plants is an odd-shaped growth that doesn't seem to belong. It is a gall that results from a tiny (1/4 inch) cynipid wasp (family Cynipidae) laying her eggs in the stem of the plant. The plant responds by growing around the disturbance, in the process creating the lumpy growth. Dozens of young wasps overwinter in the safety of the gall. Come spring they chew their way out and walk (not fly!) to the new shoots.

Huge maple-shaped leaves

Large flimsy-petaled white flowers

Blooms in the second half of June

Deep red raspberrylike berries ripen in August

Prefers deciduous woods

Other names

    salmonberry

Sparky says: Early in the season gather a few young thimbleberry shoots and try this recipe: Melt 2 tablespoons butter in a pan. Add 1 medium onion. Cook over medium heat about 5 minutes until onion is soft. Add 2 cups young

Thimbleberry shoots that have been peeled and cut small. Season with salt, pepper, and your favorite seasonings. Cook 5 more minutes until tender. Melt $1/2$ cup of grated sharp cheddar cheese over the sprouts. Serve hot on toast.

The island was ablaze in purple. Charred stumps peeked above the blooming Fireweed, not so subtly hinting at its recent and fiery past. It was July and I was on my first extended Boundary Waters Canoe Area Wilderness canoe trip with an alert guide who put all the clues together: Fireweed grows where fire has recently passed, and fire is conducive to blueberry plant growth, and July is when blueberries ripen. We stopped and picked two gallons of the fat purple berries from amongst the fireweed and charred logs. In less than an hour we had enough blueberries for three pies, a double batch of bannock, and pancakes for the next morning. Treat the large clumps of Fireweed as a purple stop sign for good blueberry picking. Its bloom, from about July 1 until August 15, coincides directly with the ripe blueberry season.

Fireweed can grow to a tall man's height, but it is often only waist high. Abundant in recent burns, often forming a dense ground cover, it also grows singly in sunny disturbed areas such as portages and campsites. It is a vertical plant with long narrow alternate leaves topped off by a spike of quarter-sized rose pink to lavender to purple flowers. The flowers bloom from the bottom of the spike up, so a single plant may have flower buds on top, blossoms in the middle, and newly formed slender seedpods on the bottom.

First responder

After a major wildfire, Fireweed serves double duty in restoring scarred and charred land. Foremost, it, along with Large-leaved Aster and Fringed Bindweed serves as the major ground cover, thereby stabilizing the soil and preventing severe erosion. Secondly, it beautifies the area with its long bloom season. Canada's Yukon Territory, which has many wildland fires, has honored the abundant fireweed as its official floral emblem. Ironically, fireweed doesn't need wild country to grow; hectares of it invaded the bombed-out areas of London during World War II.

Fireweed
Epilobium angustifolium

Tall, late-summer flower

Grows profusely in recent burns

Four-petaled purple flowers bloom from the bottom of the spike up

Bloom starts in early July and continues to mid-August

Other names

    purple firetop

    bloomin' sally

    l'herbe fret (to the voyageurs)

    rallarros ("railroad-worker's rose" in Swedish)

113

Hot honey

Connoisseurs highly regard the honey that bees manufacture from Fireweed nectar. They'll even go so far as to move the hives adjacent to a recent burn.

Whip the whoop

An infusion of Fireweed roots and leaves creates an antispasmodic tea that was used to treat whooping cough, asthma, and the hiccups.

The name Fireweed may refer to its abundance on burned-over land or it may allude to the blooming plant's resemblance to flames and the smokelike wispy down of the seed pod.

Fur trader salad

The great Arctic explorer Sir John Franklin wrote in 1823 that "the young leaves, under the name of 'l'herbe fret', are used by the Canadian voyageurs as a pot-herb." French Canadians also knew the plant as *asperge* (asparagus), for the young shoots make an excellent asparagus substitute.

The purple magnet

"When several feet tall, the plants sprouted brilliant scarlet inflorescences that shone like flames above the green patch. The flowers attracted sphinx moths at night. In the daytime they were visited by the Red Admiral butterfly, by blues, hairstreaks, Tiger Swallowtails, and others. Solitary metallic-green bees took pollen, and the Ruby-throated Hummingbird hovered to take nectar with its tongue, and most common of all were the furry bumblebees foraging on both pollen and nectar."

In a Patch of Fireweed
Bernd Heinrich

Sparky says: Test the "large Fireweed patch equals ripe blueberries" theory on your next July or August canoe trip.

Common Evening-Primrose is not a rose of the family Rosaceae nor a primrose of the family Primulaceae but, along with Fireweed, a member of the Onagraceae, a family of largely showy flowers with parts in fours (four petals, four sepals, four-parted stigma, four or eight stamens). A biennial of dry, open sites, this plant in its first year manages to push out a ground-hugging basal rosette of leaves with light, almost pinkish veins. The following year a mostly unbranched stalk shoots up 1 to 5 feet bearing untoothed lance-shaped alternate leaves and a head of flower buds on top. Buds bloom from the bottom up, usually in pairs, so that at some point in the summer, a plant may have seedpods, flowers (up to 2 inches across), and flower buds simultaneously. Yellow petals open fully at dusk to reveal eight yellow stamens and a distinctive four-branched stigma that forms an X or cross, inviting nocturnal moths, especially sphinx moths (family Sphingidae), to feed and, in the process, perform their pollination magic. Though the Common Evening-Primrose tends to close again by midday, later in the season, flowers will stay open all day.

Introduced to Europe from America as early as 1614, the evening-primrose was prized for its beauty in the garden as well as its delectability on the plate. English gardeners cultivated it for the pink taproot of the basal rosette, which when gathered early, peeled, and boiled 20 to 30 minutes in two changes of water is reminiscent of parsnips. Interestingly, Germans who ate the root and the peeled young leaves turned around and began marketing this native American plant back to the states under the title "German rampion" (rampion is a common name of the European Bellflower, (*Campanula rapunculoides*), the roots of which were eaten with the leaves in salad).

Today, evening-primroses are still a popular addition to flower gardens, but their nutritious edible qualities are largely ignored.

Common Evening-Primrose
Oenothera biennis

Showy four-petaled yellow flowers on tall stalk, 1 to 4 feet

Pioneer of disturbed soils, crevices in bedrock, and eroded banks

Blooms July through August, fully opening at dusk

Other names

    evening star

    king's cure-all

    night willow herb

    German rampion

    nattjus ("night light" in Swedish)

Miracle plant

Oil of Evening-Primrose (oil of EP) is being touted as a wonder drug not only by New Age herbalists but also by the world's medical community. In a 1981 study, St. Thomas's Hospital in London treated sixty-five women suffering from PMS with oil of EP. An astonishing 61 percent experienced complete relief, with an additional 23 percent realizing partial relief. One symptom, breast engorgement, was substantially reduced.

In November 1982, the distinguished British medical journal, *The Lancet*, presented the results of a double-blind crossover study on ninety-nine patients with ectopic eczema (severe skin rash with lesions). All were treated with high doses of oil of EP, and 43 percent experienced significant improvement of their condition. And the list of potential uses for oil of EP goes on and on:

- Calms hyperactive children in two-thirds of patients.
- Can help alcohol withdrawal, help prevent hangovers, and ease postdrinking depression.
- A study in Inverness, Scotland, has made a possible link between the oil and regeneration of liver tissue damaged by alcohol.
- Helps those who suffer from dry eyes and brittle nails.
- Combined with zinc, it is used to treat acne.
- A New York hospital gave oil of EP to a study group of people over 10 percent above their ideal weight, resulting in their losing weight.
- Contains an anticlotting factor for prevention of heart attacks caused by thrombosis (a blood vessel blocked by a clot).
- Sixty percent of rheumatoid arthritis patients at the Glasgow Royal Infirmary were able to stop their normal antiarthritis drugs after being treated with oil of EP. A treatment of oil of EP combined with fish oil was even more effective.

What is the explanation for these miraculous results from a humble northern wildflower? The answer is in the essential fatty acids, especially gammalinoleic acid (GLA), found in the extracted oil. GLA is the forerunner of a hormonelike substance called PGEI, which affects the body's internal mechanics in many positive ways. GLA may aid in the production of PGEI, which can be blocked in some people. Consult a doctor for usage and dosage of oil of EP.

 Sparky says: Plant a floral clock. Evening-primroses are dusk bloomers, but each species of flower has its own bloom chronology. Linnaeus, the Swedish botanist, made one in 1748 that was so effective that "his friends said they could tell the time, on entering the garden, from the scent of the flowers that marked the hours." Of course, not all varieties will grow in all locations, so do your homework and have some fun. Make it a family project. Here is an example.

A.M.	5–6	dwarf morning glories, wild roses, and pumpkin flowers open
	7–8	dandelions open
	8–9	African daisies open
	9–10	pumpkin flowers close, gentians open
	10–11	California poppies and tulips open
	11–12	moonflowers close
P.M.	Noon	morning glories close, goatsbeard and chicory open
	4	four-o-clocks open
	4–5	California poppies close
	sunset	evening-primroses and moonflowers open
	8–9	day lilies and dandelions close
	9–10	flowering tobacco opens
	10–2	night-blooming cereus opens

Bunchberry

Cornus canadensis

Low, woody plant with six leaves

Four white petal-like bracts surround a cluster of tiny greenish flowers

Thrives in deep shade of moss carpets in coniferous forests

Blooms late May to late June

Bunch of plump red berries by mid-July

Other names

    Canada dogwood

    dwarf cornel

    low cornel

    pudding berry

    cracker berry

Bunchberry, sphagnum moss, Clintonia, False Lily-of-the-Valley, and Starflower are the primary components of the "Canadian carpet" that underlies much of the boreal forest. These species are the few, the tough, the proud—able to survive the highly acidic conditions of the sphagnum/Schreber's Moss (*Pleurozium schreberi*) mat found under the North's Black Spruce and Jack Pine. Growing as numerous clones from a woody rhizome, genetically identical colonies may cover several square yards. A solid carpet of white-blossomed or red-berried Bunchberry creates a wonderful mosaic.

A deceiving plant, Bunchberry on the surface appears to be a small, four-petaled herbaceous flower that forms sweet red berries. In actuality, it is a plant in the same genus as the much larger and very familiar shrub, Red Osier Dogwood, with identical leaf venation. The four white "petals" are actually bracts surrounding a cluster of twenty to thirty tiny greenish flowers. Each flower consists of four minute sepals, petals, and stamens. The charade continues as these ripen into plump red berries that appear juicy but are actually quite mealy. They are not poisonous, however, and to me, sucking on a mouthful is a refreshing treat, though I have not made many converts. Large seeds deter some, while alleged tastelessness leaves others cold.

Flowering plants display six spreading alternate leaves low about the stem, which give the illusion of being a perfect whorl. One pair is normal-sized, but the other two pairs are about half-sized. Note the number of leaves on plants with flowers or berries compared to those without. It seems that when the plant finally blossoms it actually grows two additional leaves. Could it be that the energy demands of flowering require the increased photosynthetic output of the extra leaves?

During the last week in May, after the bracts turn from greenish to white, watch for the central cluster of flowers to open. The bloom lasts through mid- to late June. (Be aware, though, that a few plants have a second bloom season, which runs from August into late September, but the lateness of the season prevents the formation of fruit.) The green developing ovaries ripen into plump $^1/4$- to $^3/8$-inch red globular fruits by mid-July, with peak abundance in mid-August. Botanically speaking, the berry is not really a berry but a drupe. Drupes are a simple, fleshy fruit like berries, but unlike them, drupes have only one stony seed. Cherries, plums, olives, and peaches are all examples of drupes.

British admiration

The passionate gardeners of England, always eager to introduce something new and exotic to their flower beds, imported the Bunchberry from America in 1774. Its reputation as a colorful, shade-loving perennial spread across the British Isles, making it a most popular "alien." This growing adoration culminated in 1937, when the small dogwood received the Royal Horticultural Society Award of Merit. In America, Bunchberry is largely ignored as a cultivated plant.

Sparky says: Scottish immigrants to America called Bunchberry the "plant of gluttony" as the berries seemed to stimulate the appetite. I've never noticed this phenomenon, but decide for yourself as to their edible qualities. Grab a cluster of the ripe red berries and pop them in your mouth. Suck on them. Spit the seeds out.

I've heard that adding lemon juice can enhance their flavor, but E.L. Palmer scoffs at this, saying, "It would seem that many superior fruits could be found wherever lemon juice is available."

You can also try boiling them into a pudding and adding a cream topping, as former Quetico naturalist Shan Walshe recommended.

Bird snacks

American Robins, Veerys, and Philadelphia and Warbling Vireos all feast on the fruit while White-tailed Deer will nibble on the leaves. Nashville Warblers occasionally nest on the ground under a dense ceiling of Bunchberry leaves.

Eastern Dwarf Mistletoe

Arceuthobium pusillum

Perennial, parasitic flowering plant

Its presence is made known by a large round cluster of branches in Black Spruce

Eastern Dwarf Mistletoe is not the commercially harvested plant that hangs over your doorway at Christmas, giving one license to kiss. Rather it is a tiny parasitic plant that attacks Black Spruce and is rarely seen. What is seen is a "witches' broom," an unusually dense round mass of branches that can attain diameters of 3 to 10 feet. This abnormal growth of malformed branches is stimulated by Dwarf Mistletoe infection. It is the major cause of mortality in Black Spruce forests of the Great Lakes region and southern Canada, with 15 percent of all stands infected. Trees hosting mistletoe suffer from slower growth, poor seed production, and inferior wood quality, although it may take many years to actually kill the tree. Rare in White Spruce (except along the Maine coast), Dwarf Mistletoe occasionally attacks Red Spruce (*Picea rubens*) and Tamarack.

Explosive seed cannons

How does Dwarf Mistletoe spread? It simply shoots its seeds—up to 54 feet in fact! The fruit explosively ejects its seed at an initial velocity of over 62 miles per hour. An extremely sticky coating ensures that the seed will adhere to any branch in its path—preferably the branch of a Black Spruce. The odds are pretty good, though, as Black Spruce tends to grow in single-species stands. But a mystery remains. How does one explain the occurrence of Dwarf Mistletoe on islands off the Maine coast and in Lake Michigan, nineteen miles from the nearest infection?

Critter power

Forest researchers Michael Ostry, Thomas Nichols, and D. W. French set out to answer the question of long-range mistletoe dispersal. They set their sights on animal vectors, though it was well known that the seeds are unpalatable to birds and mammals. In the literature they found that the sticky seeds had been discovered on the feathers of Gray Jays, Black-capped Chickadees, Red-breasted Nuthatches, and American Robins and in the fur of Red Squirrels.

In their own investigations, the researchers netted birds and mammals in an infected Black Spruce stand near Cloquet, Minnesota, during the peak seed dispersal time, mid-September. In 1974, 12 percent of all critters caught had at least one mistletoe seed stuck to them. This figure jumped to 20 percent in the 1975 study. Critters acting as seed dispersers were Gray Jays, migrating warblers, Red Squirrels, and Northern Flying Squirrels. Evidently, as animals forage amongst the witches' broom they trigger the explosive ejection of seeds. The gooey projectile sticks fast to any animal in its path, which now becomes a vector, spreading the parasite to wherever it stops to preen the seed out of fur or feathers. One Dark-eyed Junco stuck with three hitchhiking seeds was captured 155 yards from the nearest infection center. Boreal Chickadees were the champion broom foragers, and red squirrels were the most frequent utilizers of the brooms for nests, but the researchers' conclusion was that migrating birds are most likely responsible for the long-range dispersal of Eastern Dwarf Mistletoe.

Sparky says: Plants get around. And the way they get around is through seed dispersal. Dwarf Mistletoe and Spotted Touch-me-not (*Impatiens capensis*) catapult their seeds. Dandelions (*Taraxacum officinale*) and Sugar Maples let the wind give their seed pods a ride. Plants with bright-colored and juicy berries attract critters who devour them en masse and poop out the seeds elsewhere. And then there are the "hitchhikers." The dried seedpods of these guys are armed with special hooks and barbs that snag any passerby, whether deer, dog, or your dress pair of Dockers. They travel thus until preened out, and if they like the soil, they put down roots. Plants we call weeds often utilize this tactic.

To explore further the world of "hitchhiking," try this family autumn activity. Find a field or vacant lot where a variety of plants have mostly gone to seed. Late September and

October are best. Now put a big pair of old wool socks over your shoes and frolic in that field. Play tag, hide-and-seek, whatever, but run and play until you have a load of hitchhikers stuck to your socks. Gather everyone around and examine the different types of seeds. What mechanism did they use to attach to you? What plant did they come from?

Take it one step further and plant your socks! Each person lays his or her socks flat in a pot of dirt. Cover them with an inch or less of soil. Keep the pots outside over the winter— many plants need that cold period of dormancy. You'll be amazed at what grows from your "soiled laundry."

Wild Sarsaparilla
Aralia nudicaulis

Wild Sarsaparilla is a perennial growing from an aromatic underground rhizome. The leaf stem and flowering stalk appear to arise from the soil independently but actually sprout together from a woody stem just below the surface. The "leaves" are actually a single leaf whose petiole is divided into three divisions, each with three to five leaflets. It is one of the first signs of life in the North Woods spring. It doesn't emerge green, but rather a deep burgundy. The leaf stays this color almost until reaching its full 12 inches. The veins are the first part to turn green.

Flowers bloom just as the leaves turn from burgundy to green. The tiny white flowers form three perfectly round heads (umbels), often found just below the overhanging canopy of the leaf. Late May to mid-June is the bloom period. Look for them in wooded uplands, especially under deciduous trees.

Berries are large relative to the flower size. They ripen to a deep purple-black in mid-July. Red Foxes and Black Bears eat them, but humans steer clear.

The roots of root beer

The legend goes that a new drink was introduced into the lumber camps of the North sometime around the turn of the century. Its name was "sarsaparilla," and its reception by the lumberjacks was lukewarm. Undaunted, the entrepreneurs tried a new marketing strategy and renamed the beverage "root beer"! The rough-hewn loggers saw this drink in a new light, and it became a success overnight.

Whether the root in "root beer" was Wild Sarsaparilla or not remains to be seen. It does have an aromatic root (like ginseng and most other members of the family Araliaceae), which has been used as a substitute for the true sarsaparilla, *Smilax officinalis*, which grows in Honduras.

Ojibwa uses

The Ojibwa Indians of the Lake Superior region had all kinds

Abundant herb throughout the canoe country

Doubly compound leaf arising separately from flowering stalk

Tiny white flowers and blue-black berries held in three spherical clusters

Blooms in late May to mid-June

Berries ripe by second half of July

Other names

   shot-bush

   rabbit's-root

   wild licorice

   false sarsaparilla

The Root Beer Lady

Dorothy Molter, the late famed "Root Beer Lady" of Minnesota's Boundary Waters, served homemade root beer to thousands of trail-thirsty paddlers every summer. It wasn't as cold nor as tasty as the stuff you could get out of the pop machine for four bits but, boy, it was the best thing going on the Canadian border, one hundred miles north of nowhere!

of uses for Wild Sarsaparilla root. Skin sores were soothed by a poultice of mashed root. To stop a nosebleed they placed a chewed fresh root in the nostril. A decoction of sarsaparilla root, Spikenard (*Aralia racemosa*), and Red Swamp Currant (*Ribes triste*) was effective in stopping excessive menstrual bleeding. When your horse gives out and is about to drop, rub a mixture of root steeped with other herbs on its chest and legs. The human circulatory system is also rejuvenated by a root decoction taken internally.

To make a charm effective in luring fish, the Ojibwa rubbed their nets with a mixture of Wild Sarsaparilla root and Sweet Flag (*Acorus calamus*) root.

 Sparky says: Now, I don't recommend making the drink sarsaparilla or root beer from our Wild Sarsaparilla roots, but why not take a shot at making it from root beer concentrate that's available at any grocery store. Here's the recipe for a 24-quart batch.

Mix one 2-ounce bottle of concentrate with 5 pounds of sugar in a large nonaluminum pot or crock. Slowly stir in 5 gallons of lukewarm distilled water. Dissolve 1 packet of yeast (1/4 oz.) in 1 cup of warm water. Add to root beer mix and stir. Immediately pour into hot sterilized bottles, leaving a 2-inch head space. Cap or cork tightly. Do not use plastic bottles or twist-off caps.

Store in a warm place (70° to 80° Fahrenheit) such as on top of the refrigerator for ten days or more. Keep bottles on their sides. Chill before serving.

This process is not a sure bet, so you may want to halve the recipe for your first batch. Also remember that the flavor will be unique. It won't taste like A & W!

I must admit that Harebells are one of my favorite plants. Their delicate stems and large blue flowers contrast so nicely with the orange lichen-splattered rocks and cliffs they inhabit. Quetico naturalist Shan Walshe found them more abundant on metasediments and greenstone than on granite. Paddling along bare rock shores is the best way to find and admire them. They often grow in clumps, isolated from other flowers by the harsh realities of life on the rocks.

These are the famed "bluebells of Scotland" that poets have waxed eloquently about for centuries. In fact, the Harebell is a very cosmopolitan plant, inhabiting rocky crags from Norway's Sunndalsfjorden to Russia's Kamchatka Peninsula, south to Asia, and west to the canoe country of Minnesota and Ontario, the Great Lakes, and New England. Actually this get-around plant reaches elevations of 5,000 feet in all the high western mountain ranges except the Sierra Nevada.

The grasslike stems with sparse slender leaves grow to a height of 1 foot. Don't be confused by the Latin specific epithet, *rotundifolia*, which refers to the tuft of round, sparingly toothed leaves in spring that wither and die before the plant flowers. Lavender bell-shaped flowers 3/4 inch wide hang nodding from their tips. The corolla is five-lobed, bearing inside a prominent pistil and five anthers. Bumblebees, though about the same size as the blossom, are the chief pollinators. In its awkward attempts to gain entry to this inverted bell, the bee grasps the stigma and in its struggles is brushed with pollen. Bumbling off to the next flower, the bee passes the pollen on to its stigma—a sexual mission accomplished by an ignorant accomplice.

Though most Harebells are blue, I once chanced into a rare clump of white-blossomed Harebells growing 200 feet above Lake Superior on the vertical cliff face of Minnesota's Palisade Head.

Harebells are in bloom from the third week of July until the end of August.

Harebell
Campanula rotundifolia

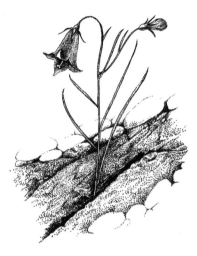

Common in crevices of bare bedrock

Hairlike stems and leaves

6 to 18 inches tall

Blue-purple inch-long bells hang from tips of stems

Bloom from mid-July to mid-August

Other names

    bluebells

    bluebells of Scotland

    Scotch bellflower

    lady's thimble

    witch's thimble

    ziginice ("pouring" in Ojibwa)

    blålokke ("bluebell" in Norwegian)

The thimble of witches

Our ancestors were poor spellers. At least that is one theory behind the origin of the vernacular name "harebell," which is said to refer to the hairlike slender stem leaves. Or maybe they could spell, and the name alludes to an association with witches, who in colonial America were believed to be able to transmute to hare form, a bad-luck animal when it crossed your path. In fact, an old name for this plant in Scotland was "witch's thimble."

Sparky says: Make yourself a small plant press. I made one a few years ago that fits nicely into my daypack. When I come across a group of unfamiliar flowers, I gather up a single specimen and press it for further study at home. Start by cutting out two 5-inch squares of stout plywood (preferably ⁵/₈ inch thick). Drill a hole in each corner ¹/₂ inch in from the edge. Join the two pieces with four 3-inch bolts and wing nuts. Before going into the field, layer 5-inch squares of corrugated cardboard with the corners nipped off between the boards. Between two pieces of cardboard place two sheets of blotting paper. After finding a suitable plant specimen, take off the top board of your press and sandwich the flower between the bottommost pair of blotting paper sheets. Make sure to position it so stamens and other key identifying characteristics can be seen. Carefully replace the top and screw together tightly. Your press should be able to accommodate three or four pressing layers. After returning home, let the press sit tightly screwed down for several days. You can preserve specimens by placing them between clear contact paper and a stiff white cardboard. Key the plant out, label, and file. You've just started your own herbarium!

The first week in June is when you may find your first delicate pink blossoms of the Twinflower deep in the shady, moss-covered recesses of the forest. Searching out this tiny, fragrant wildflower is well worth the effort, but to really appreciate it, you must get down to its level and look and smell. Blooms may last into July, with a possible second bloom in mid-August.

The Twinflower is appreciated worldwide growing in the boreal forests of Scandinavia, Finland, Russia, and our own North America.

The Twinflower is the smallest member of the honeysuckle family, with upright flowers standing only as tall as your pinkie finger. Two nodding pink bells hang from every flower stalk. A pair of rounded small leaves originate at the stalk base. This is not the entire plant, though, for many flower stalks arise from an above-ground creeping stem that may reach 3 feet in length. The overall effect is that of a Lilliputian boulevard lined with pink street lamps.

"Twinflowerus latinus"

Twinflower also grows in Sweden, the home of Karl von Linné (1707–78), naturalist, botanist, taxonomist, and father of systematic biological classification. The Linnaean system of Latin binomial nomenclature, in which each species possesses a unique genus-species name, is used for all plants in this book and for all organisms worldwide. This ordered system allows scientists from every corner of the earth to communicate clearly about specific species using their universal Latin name. Linné was so into Latin naming that he gave himself one, changing his name to Carolus Linnaeus.

A student of Aristotle's, Theophrastus (370–285 B.C.), first separated the trees from the shrubs from the herbaceous wildflowers. French botanist Charles L'Ecluse gave plants two names as early as 1576, but there was no consistent system. Publication of Linnaeus's *Genera Plantarum* in 1753 introduced organization of the plant kingdom by sexual characteristics (number

Twinflower
Linnaea borealis

Two delicate, pink bells nod from a 2-inch stem

Many flowering stalks arise from a creeping stem

A pair of small round basal leaves

Found in mossy coniferous woods

Blooms in June

Other names

  Linne'a (in Norway and Sweden)

of stamens and pistils, etc.). The system used today is little changed from the one he put in place 250 years ago.

Born the son of a minister in Rashult, Sweden, young Linnaeus was nudged toward the ministry. Fortuitously, he ended up in medical school, where, in that day, botany was the principal requirement. It was then he started his first gardens, which still attract hordes of visitors to Uppsala. His "herborising lectures" were the social events of the summer. Linnaeus would wander through the Swedish countryside ahead of two hundred to three hundred flower enthusiasts, accompanied by a band of trumpets and French horns that played when he wanted to share some information on a topic.

He was considered by many to be strange yet brilliant, brash, and even arrogant by some standards. For example, he collected tales of men who had deceived their wives and in return had been deceived by their wives. This he correlated to the sexual trickery that goes on between insect pollinator and plant, especially the orchids. Linnaeus once wrote that "the genitalia of plants we regard with delight, of animals with abomination, and of ourselves with strange thoughts." He believed that God had given him the mission to bring order out of chaos by classifying and naming all living things, and this he did according to the sexual characteristics of plants.

His brashness showed when he asked his former teacher, J.F. Gronovius, to name the Twinflower after himself, a practice unheard of by honorable botanists. Linnaeus fancied that the flower was much like himself, "lowly, insignificant [and] flowering for a brief space." And it was his favorite plant—he even held a sprig in his official portrait. It seems a fitting tribute to our greatest taxonomist that Twinflower should be forever known as *Linnaea borealis*.

National treasure

Linnaeus is a national treasure in Sweden. Everywhere you go, his contented countenance shines. A larger-than-life statue sits in a Stockholm park, where he clutches, not the delicate

Twinflower, but rather a bunch of flowers that appear to me to be from the primrose family. Every time Swedes pull 100 kronor bills from their pockets they encounter the great botanist. And in Uppsala's *domkyrka* cathedral, where he is buried alongside his wife and child, his chiseled porphyry silhouette is labeled with the Latin title *princeps botanicorum*, appropriately "The Prince of Botanists."

 Sparky says: Get down on your hands and knees and gently pull the creeping stem of Twinflower from the sphagnum moss to get a better feel for how this plant spreads, being careful not to pull up any roots. How long is the creeping stem? Who can find the longest one? How many flowering stalks arise from one creeping stem? Replace the plants carefully.

Part two: Accept the Linnaean challenge and memorize the Latin names (Genus-specific epithet) for ten of your favorite plants. Write the common name ("Twinflower") on one side of an index card and the binomial Latin name (*Linnaea borealis*) on the other side. Quiz each other.

Large-leaved Aster

Aster macrophyllus

Very large heart-shaped basal leaves can cover the ground in certain areas

In second year, tall flower stalk shoots up

Many small purple-rayed, yellow-centered daisylike flowers cluster at top

Usually the last flower blooming in the North Woods (late July to early October)

Other names

  lumberjack toilet paper

It is early August in the Boundary Waters Canoe Area Wilderness, and the first purple blossoms of the Large-leaved Aster dot the portages. Spurting up from its basal leaves in late summer, this aster is easily noticed blooming at waist height, high above other plants. But the Large-leaved Aster can be just as recognizable in its vegetative state. Early in spring, the large heart-shaped basal leaves emerge to carpet many parts of the northern forests, especially in the richer soil beneath deciduous Quaking Aspens. These leaves are big enough to have covered Adam, had the canoe country been the Garden of Eden (some say it is paradise!).

The inch-across purple and yellow blossoms appear to be one flower but actually are many flowers. The ten to sixteen pale purple "petals" are actually ray flowers, while the yellow centers are many tiny disk florets. Get down close and see for yourself. Each tiny floret possesses its own stamens, pistil, and ovary. This feature makes all asters members of the Compositae family, which also includes the dandelions, sunflowers, and daisies.

Bumbling bees

A friend of mine questioned me the other day on some odd bumblebee "bee-havior" she had observed this fall. Apparently the bees, after feeding on Large-leaved Aster flowers, were falling to the ground. Alive, but obviously floundering, they were unable to fly. Inebriated? Careless (after all they are "bumble" bees)? No, none of these answers fit. The answer came while I was reading entomologist Bernd Heinrich's classic *In a Patch of Fireweed*. He observed a certain species of bumblebee *Bombus terricola* on cool New England mornings foraging on high-nectar, high-reward flowers such as Spotted Touch-me-not, Fireweed, and milkweed. But only on warm days were the bees ever seen foraging on the low-nectar, low-reward plants such as Large-leaved Aster and goldenrod. Heinrich's conclusion was that "bees seemed able to calculate profits and losses and to behave appropriately. When flowers with ample nectar (Fireweed, jewelweed, milkweed) are available, the bees tend to stay away from less-rewarding flowers (aster, golden-rod), especially at low air temperatures where they have to expend more calories to keep warm (shivering) than they can get back from the flowers." So, the falling-down bees my friend observed were in an energy Catch-22. The cold weather and limited flower selection of September meant that the bumble-bees, if they were going to forage at all, had to feed on the low-reward Large-leaved Asters. Here they walked around the flowers feeding as their muscles cooled. Unable to take in enough energy to raise their body temperature to 86° Fahrenheit, they found flight impossible and hence tumbled to the ground. Flight on that September day would have required either warmer temperatures or high-reward flowers to forage on. Thanks, Bernd. Case closed.

Lumberjack toilet paper

Many a woodsman and woodswoman has resorted to using the broad basal leaf of the Large-leaved Aster as toilet paper in times of need. The bottom of the leaf is rough and hairy, but it satisfies the most critical criterion: surface area.

Sparky says: Large-leaved Aster grows around nearly every latrine and outhouse in the North Woods. This knowledge can come in handy, wiping away all fears of ever being caught without real toilet paper. Why not then, when on your next camping trip, leave the TP behind and sample mother nature's best wipe, aster leaves.

Large-leaved "star"?

The word *aster* is Greek for star and is the root of asteroid and asterisk (*).

Spotted Joe-Pye Weed

Eupatorium maculatum

4- to 7-foot-tall purple-stemmed plant with whorls of four to five leaves

Flat-topped cluster of tiny "fuzzy" magenta flowers at top

Grows in marshy edges and other very wet places

Blooms from mid-July to the end of August

Other names

   boneset

   kidneyroot

   motherwort

   king-of-the-meadow

Spotted Joe-Pye Weed, occasionally lofting to 7 feet from soggy roots to topmost shoots, is the Empire State Building of the herbaceous plant world. There are a handful taller, but none more beauteous. And its height is crucial to its survival, for come late July and August, when this latecomer flowers, it has to reach heights above and beyond the surrounding vegetation to wave its magenta flag for the insect world to see. The sturdy purple or purple-spotted stem is encompassed all the way up by whorls of four or five pointed leaves with saw-toothed edges. Leaves can reach 8 inches long or more. Purple, pink, or magenta flat-topped fuzzy flower clusters are located terminally on the stalk.

Getting his feet wet is no problem for old Joe-Pye, in fact he prefers it, thank you. Beaver meadows, marsh edges, rivulet borders, portages, floodplains, and Black Ash–Northern White Cedar swamps is where Joe prefers to make his home. But a field biologist friend of mine discovered that Joe might not like such soggy feet after all. He noted that the plant preferred the slightly drier microhabitat of very old beaver dams. The dams were so decayed as to be unnoticeable except for the line of Joe-Pye growing along their lengths. Look for blooming specimens in the canoe country as early as July 10, but peak is not until August.

Who was Joe Pye?

According to legend, Joe Pye was a Native American doctor, or herb healer, who treated rural "white folk" sometime between Pilgrim days and colonial times. He is credited with stemming a typhus plague by using one species of *Eupatorium*. It is not clear if this credit came from actual documented sources or claims by Joe himself. He did claim to "set shaking bones to rest in ague-rent bodies." This may be the same malady, known as "break-bone fever," a virulent form of flu, that was common in the southern colonies at the same time. It was treated with a tea of *Eupatorium perfoliatum* leaves and is most likely the source of the common name "boneset," which has also been applied to several other species of the genus *Eupatorium.* Unfortunately, the truth of Joe Pye and his miracle cures may be lost forever in the beaver meadows of time.

Today, herbalists do use the rhizome and roots of a very closely related species, Sweet Joe-Pye Weed (*E. purpureum*), as a diuretic to treat urinary infections and stones. It has also been effective in toning the reproductive system, treating inflammation of the prostate, preventing menstrual cramping, and encouraging excretion of excess uric acid and so thereby treating rheumatism and gout. Constituents of the plant include a volatile oil, resin, and flavonoids (including eupatorin) that, according to M. D. Midge and A. V. Rao in their 1975 *Indian Journal of Chemistry* article, may possess anticancer properties.

Antidote

The genus name *Eupatorium* commemorates Mithridates Eupator, King of Pontus (an ancient country of northeast Asia Minor from the 4th century B.C. until 66 B.C.), and his discovery that one species was an effective antidote to poison. What poison and what species remain a mystery.

Sparky says: Aliens! Alien species surround us everyday. Dandelions, European Starlings, House Sparrows, and Rock Doves thrive in the wild in their "New World." Dandelions were brought from England by health-conscious Pilgrims, who grew the vitamin C-rich green in their gardens. European Starlings were introduced to New York's Central Park by a Shakespeare fanatic who wanted to be surrounded by all the birds mentioned in the master's plays. Even our pristine North Woods has been invaded by Old World plants. Learn the identity of an alien in your neighborhood and research its origins. How did it get here? How did it spread? Has it become a problem?

Broad-leaved Arrowhead

Sagittaria latifolia

Aquatic plant of lake and river edges

Arrowhead-shaped leaves are distinct

Leaves can be narrow or wide

Three-petaled white flowers are in whorls of three on separate emergent stalk

Blooms in late July to mid-August

Other names

   wapato (Chinook)

   duck potato

   swan potato

   bredpilblad ("wide pill-leaf" in Swedish)

Out of the water along the lake's margin grows a hard-to-miss plant with distinctive spear- or arrow-shaped leaves. The leaves are three-pointed and can be broad or quite narrow. A separate stalk bears the flowers, which have three round white petals and fuzzy yellow centers. Whorls of three flowers ring the stem. Look for it to be in bloom from the third week in July to mid-August. The resulting fruits are round green heads.

The magnificent wapato

The real story of this plant is found below the waterline. Along the mud-buried roots are found potato-like starchy tubers that Native Americans, Muskrats, and ducks have used as food for centuries. Sixteen species of ducks and geese are known to feed on the fruit and tubers. Tubers vary from new-potato size up to those rivaling the cultivated white potato. One narrative describes the collecting process a Native American woman used. Wading chest deep in the water, she "by means of her toes, separates from the root this bulb, which on being freed from the mud, rises immediately to the surface of the water and is thrown into the canoe."

To prepare this starchy delight, simply wash off, then either boil, fry, or hash. Anything that can be done to a potato can be done to the arrowhead tuber. Nineteenth-century German explorer-ethnologist Johann Kohl describes dried "swan potatoes" strung on bois blanc (cedar) hanging in the lodges of the Lake Superior Ojibwa. Once dried, they could easily be grabbed by the handful and thrown into the boiling pot. He continues that "when the Indian squaw described her delicacies to me, my mouth always began to water, such exquisite qualities did she give them."

Raw sewage treatment

The latest thing in small municipal sewage treatment is human-made wetlands, or as *Audubon* magazine titled it, "plants that purify." Benton, Louisiana, a town of 2,500 folks, has created a twelve-acre sewage treatment marsh. Also known as microbial rock-reed filters, these wetlands and their plants shelter microorganisms that consume ammonia, nitrogen, and phosphorous and help break down detergents, chemical pollutants, and pesticides into simple compounds that can be absorbed by the plants. Main players in the game are African Calla Lilies, aquatic irises, mini and giant bulrushes, and arrowheads, of course. Each species' chemical secretions allow specific microbes to coexist with it and carry out part of the breakdown mission. Arrowhead's specialty is ammonia decomposition.

Built in 1987, Benton's twelve-acre treatment marsh can process 200,000 gallons of raw sewage per day, all with plant-power. But it takes one to three months for water to trickle through the system, whereas it would take a mere forty-eight hours in a conventional plant. The advantages, though, are no odor, no chemicals, no cement or iron, great wildlife habitat, and no leftover sludge (which contains the heavy metals that in most systems are landfilled or incinerated). Large land requirements and cold microbe- and plant-slowing winters are the limiting factors to broadscale implementation of natural sewage treatment. But it does work. Citizens of Benton get their drinking water 3 miles DOWNstream from the plant. Now that's a testimony.

Lewis and Clark fare

Due to its edible qualities, the Broad-leaved Arrowhead is also known as "duck potato" or "wapato," a Chinook name made popular by Lewis and Clark. While at the mouth of the Columbia River during the winter of 1805–06, Lewis, the botanist, wrote: "We purchased from the old squaw, for armbands and rings, a few wappatoo roots, on which we subsisted. They are nearly equal in flavor to the Irish potato and afford a very good substitute for bread."

 Sparky says: After you've hashed them up some "wild potatoes" for breakfast, the kids will be heartily restored and ready for some underwater shoreline research. To make a Jacques Cousteau–approved "self-contained underwater viewing apparatus" (SCUVA), simply cut the bottom out of a large empty tin can. Now stretch a heavy clear plastic bag across the bottom and fasten it in place with rubber band or duct tape. By sticking the SCUVA halfway in the water, the kids can reveal a clear picture of the underwater realm. Have them wade the shallows with their SCUVA gear looking for crayfish, caddis fly larvae, minnows, and more.

Clintonia

Clintonia borealis

6 to 16 inches tall

Two or three large parallel-veined basal leaves

Three to ten pale yellow lilylike flowers nod from top of leafless stalk

Blooms from last week in May through mid-June

Blue berries ripen by mid-July and stay blue into August

Other names

   blue-bead lily

   corn lily

   Clinton's lily

   cow tongue

Clintonia is a shade-loving plant that thrives in varied habitats. Famed naturalist Shan Walshe found it to be the dominant understory plant beneath a rich-soiled Red Maple forest at Glacier Lake and, conversely, found it to be abundant in an acidic Black Spruce–Northern White Cedar swamp near French River of Northwest Ontario's Quetico Provincial Park. Since it is a perennial reproducing from a creeping rhizome, huge mats of this plant may be found. Clintonia is a common plant throughout the North Woods.

This 6- to 16-inch-tall member of the lily family has two or three broad parallel-veined shiny basal leaves and three to ten drooping yellow flowers on a leafless stalk. In my observations, I found most mature plants to carry seven to ten flowers per stalk. Several bore a pair of flowers growing directly off the stem $1^{1}/2$ inches below the top. Flowers in the Liliaceae family have flower parts in multiples of three. Clintonia carries on the family tradition by possessing six sepals and six stamens. Watch for it to flower between the last week of May and early June. By the second week in July the ovaries have ripened into berries of rare blue—they have no tint of purple whatsoever.

Cows, corn, and Clinton

Blue-bead lily, cow tongue, and corn lily are vernacular names given to this plant. New York Governor DeWitt Clinton (1769–1828), an early naturalist, was the honoree of the bestowed genus name *Clintonia*, which also gave rise to its common name. Governor Clinton was most famous for his promotion of the construction of the Erie Canal. My letter to the White House concerning ancestral ties to current president Bill Clinton has yet to be answered.

Dr. Jekyll and Mr. Hyde

Clintonia is the Dr. Jekyll and Mr. Hyde of the edible plants. While its young cucumber-tasting leaves are a treat for people and deer, its berries are mildly poisonous to humans. (Eastern

Chipmunks can stomach them though.) Children, especially, must be warned that not every blue berry in the woods is a yummy, sweet, and juicy blueberry. One taste is usually enough to discourage most accidental pickers.

 Sparky says: If on an early canoe trip, try a Clintonia leaf salad. It makes an excellent firm and crisp lettuce. Pick only the youngest, still-curled leaves (3 to 4 inches tall). As the leaves age they acquire a bitter aftertaste. Make sure they are Clintonia leaves, and obey the "wild grazers' creed" by leaving nine out of ten plants in the area intact. Bring a packet of your favorite salad dressing to top it off. I recommend French.

Rose Twisted Stalk

Streptopus roseus

Zigzag stem arches out 12 to 24 inches

Six-pointed pink bells hang singly from leaf axils

Parallel-veined leaves are alternate

Fruit is a cherry-red translucent berry

Flowers in early June and berries in late July and August

Other names

twisted stalk

Rose Twisted Stalk grows nearly everywhere in the canoe country, but it is downright abundant in the pockets of deciduous forest. It thrives in the rich, less acidic soils. Watch for it along portages and other trails.

As a member of the lily family, twisted stalk belongs to the monocots, one of the basic divisions of angiosperms (flowering plants). Monocots have one cotyledon, parallel leaf veins, and flower parts in threes. Lilies, irises, orchids, cattails, and grasses are familiar monocots, but most plants are dicots. Dicots possess two cotyledons, netlike leaf veins, and flower parts usually in fours and fives.

Never erect, the two or three stems of Rose Twisted Stalk, arising from a common point, arch in different directions. They are 1 to 2 feet long and zigzag. The zigzags are created by the stem changing directions at each juncture with a leaf. The 4-inch-long leaves are pointed, parallel-veined, and alternate.

June 5 to 15 is peak flowering time for Rose Twisted Stalk. A single delicate, rose-pink, six-pointed bell hangs from almost every leaf axil. The ovaries of these flowers ripen into translucent, cherry-red berries by late July. "Liver berries" is what Maine woodsmen call them. Some consider them to be a "pleasant nibble," but too many will cause diarrhea. One naturalist described them as having an "insipid, cucumber-like flavor."

Sparky says: Here's a sensory scavenger hunt for the whole family.

Feel: rotten wood, a prickly plant, carpetlike moss, sunshine on your face, wind in your hair, the inside of a woodpecker hole, pollen.

See: a plant with a zigzag stem (hint: Rose Twisted Stalk), a red leaf, a leaf ten times longer than it is wide, a dead plant that supports life, one fungus with pores and one with gills, a "helicopter" or "parachute" seed and the plant it came from, a blue flower.

Smell: dried pine needles, fresh crushed spruce or cedar

needles, something with no smell, a big noseful of fresh air, dirt, something rotting.

Hear: leaves rustling, wind in the pines, your fingers scratching bark.

Write up a list with sketches and check boxes and distribute to the kids (or adults). After the allotted time gather and share your discoveries.

False Lily-of-the-Valley

Maianthemum canadense

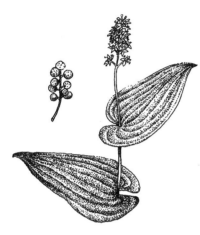

Small, common flower of shaded coniferous forests

Fragrant cluster of white flowers at top of stem

Speckled, translucent berries

Only two heart-shaped leaves

Flowers last week of May until late June

Other names

    Canada mayflower

    wild lily-of-the-valley

    squirrel-berry (Finland)

    bead ruby

A delicate and fragrant addition to the "Canadian carpet," False Lily-of-the-Valley most often occurs in colonies with Bunchberry, Twinflower, Clintonia, and Starflower on mossy beds under spruce–fir–Jack Pine canopies. False Lily-of- the-Valley is a cloning herb sprouting up plant stems at intervals along its underground creeping perennial rhizome. One-leaved plants that do not flower far outnumber two- or three-leaved flowering stalks. They are mini-solar panels, collecting sun and producing food by photosynthesis, eventually passing it down the "pipe-line" of rhizome to fuel growth of new stems.

As in all members of the lily family, leaf venation is parallel. Two heart-shaped leaves, up to 3 inches long, clasp the zigzag stem, one pointing one direction and the second one pointing the opposite direction. Atop the 3- to 6-inch stem is a starry cluster of a dozen or so diminutive white flowers ($^1\!/_8$ to $^1\!/_4$ inch across). Unlike most lilies, whose flower parts are in threes, False Lily-of-the-Valley has four stamens, two sepals, and two petals. The swept-back petals and sepals and the erect stamens give this flower its starry appearance. Blooming in late May, False Lily-of-the-Valley is true to the Greek translation of its genus name, *Maianthemum*, which means May blossom.

Bees are probably the main pollinator. The resulting pale, speckled translucent berries remain for most of the summer, ripening to red and eaten by Ruffed Grouse, Eastern Chipmunks, and White-footed Mice. Bittersweet to human palates, the berries should be eaten in moderation as they can cause diarrhea.

Sparky says: The Latin name of False Lily-of-the-Valley is *Maianthemum canadense*, which literally translates to May blossom of Canada. (This is true, but it also is a June blossom of Minnesota.) Latin names help us communicate about plants and other organisms. A botanist in Scandinavia knows False Lily-of-the-Valley by a different common name than does a naturalist in Michigan. So if they want to talk about this flower, they must first make sure they are both talking about the same species. The Latin binomial name assures that they do.

Certain Latin words pop up time and time again. It would behoove the serious naturalist to know their meaning. Look up and define the following: *virginiana, borealis, alba, nigra, odorata, rotundifolia, latifolia, purpurea, biennis, umbellata, macrophyllus, stolonifera, vulgaris, terrestris,* and *aquatica.* Once you know their definitions, many Latin names will make much more sense.

Wild Iris

Iris versicolor

Shoreline and marsh wildflower

Long swordlike leaves

Large purplish iris flowers

Bloom mid-June to July

Other names

    blue flag

    "fleur-de-lis"

    liver lily

    clajeux (Quebec)

Many garden flowers could be called gaudy, but very few wildflowers would fit that description, with the possible exception being the Wild Iris. Not gaudy in the tacky sense of the word, but in the large, bold, and colorful sense of the word.

Two to 3 feet high, the Wild Iris (or blue flag) likes to have its feet wet. The swordlike leaves and stem arise from marsh and lake edges, often from underwater. The single purplish blue flower, 3 inches across, is unmistakably an iris. What appears to be six petals, though, is actually three sepals and three petals. The lowest petal, or tongue, of the flower is patterned with yellow. It is believed that pollinating insects follow this "yellow brick road" to the nectar, only to be dusted with pollen at the "gates of Oz." At its next Oz, the unknowing insect pollinates the flower by passing the pollen to the pistil.

The Wild Iris adds color to the marshes and lake margins from the second week in June until early July. Great clumps of blooming iris are but one of the canoe country's visual treats.

The resulting seedpod is also very noticeable. It is light green and 3 inches tall, sitting vertically atop the stem.

Poison flag

CAUTION! The fresh rootstalk and leaves of Wild Iris contain iridin, a poison that can cause extreme vomiting and a bad case of diarrhea when ingested in full doses. Some people are so allergic to this substance that merely brushing a leaf may cause a skin rash. But when you dry and pulverize the root, iridin is rendered harmless. In fact, early botanist William Bartram, in 1777, described the Otasses Indians using it as a cathartic and as a poultice to treat sores and bruises. "Liver lily" was a common colloquial name for the plant, which had a reputation as a blood cleanser. Herbalists today still prescribe the dried and pulverized root as a blood purifier and remedy for skin irritations, but also claim it relieves heartburn, belching, nausea, headaches associated with digestive problems, and flatulence. Modern medicine also acknowledges that it acts on the liver and gall bladder to increase the flow of bile.

The bold and the beautiful

English essayist, art critic, and reformer John Ruskin christened the Wild Iris "the flower of chivalry," with a sword for its
leaf and a lily for its heart. Fellow nineteenth-centurian Henry
Wadsworth Longfellow echoed these sentiments by describing
it as "a flower born in the purple, to joy and pleasance."

Peace goddess

Iris was the Greek goddess
of the rainbow, who brought
peace after the gods' stormy
confrontations.

Sparky says: Visit the iris gardens of Swan Lake in
Sumter, South Carolina. The one hundred acres
of two hundred varieties and six million plants is
in its radiant rainbow-of-color peak from May 20
to June 15. Stand awestruck before the prized
Japanese Irises, 5 feet tall with blossoms 10 inches across.

Pink Ladyslipper

Cypripedium acaule

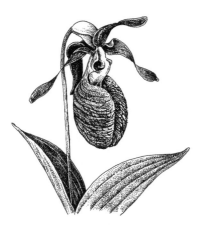

Striking pink-pouched orchid standing 6 to 18 inches tall

Two large parallel-veined basal leaves

Dry, rocky, moss-covered sites under Jack Pines and also in low Black Spruce forests

Flowers in late May to mid-June

Other names

   moccasin flower

   stemless ladyslipper

   whip-poor-will's shoes

   squirrel's shoes

   Noah's ark

   nerveroot

   old goose

Thirty-thousand orchid species dot the globe from the Arctic to the tropics, mountainside to prairie, and every habitat in between. North of Mexico, though, we only find about 177 species, with 10 of these in the genus *Cypripedium*, the ladyslippers. Six ladyslippers are found east of the Rocky Mountains, and all six are found in Minnesota. The only common ladyslipper in the canoe country is the moccasin flower, or Pink Ladyslipper. But what a commanding presence it is. At the height of blooming in late May and early June, certain locations are magnificently spattered with it. Very cosmopolitan, the Pink Ladyslipper can be as abundant in Black Spruce bog forests as on moss-covered rocky sites under Jack Pine. The common denominator is acid soils.

Twelve- to 18-inch plants are common, but I've also seen dwarf bloomers shorter than a writing pen. The only leaves are two parallel-veined basal leaves that arch backwards, occasionally nearly lying prostrate on the moss. It is the only ladyslipper with no leaves on the stem. Whip-poor-will's shoes, moccasin flower, and ladyslipper all refer to the uniquely shaped petals that form an inflated pouch. In fact, *Cypripedium* means Venus slipper in Greek. But I read recently of one grouchy botanist who felt that they were slippers "fit only for very gouty old toes." The 2-inch-long pouch is actually a modified fused petal deeply slit nearly to the bottom, heavily veined with deep pink, fading towards the bottom. The other two winglike petals are burgundy. For an interesting scent sensation, sit and sniff a flower. The faint sweet odor brings back memories of grandma's "lilac water."

The century club

For good reason, it is illegal to pick ladyslippers in many states. Seeds have an appallingly low 0.001 percent germination rate, and those making it require three years before even emerging into sunlight. Once established, a plant may bloom for one hundred consecutive summers or more! A century of survival can be plucked in one thoughtless instant. Pass the word.

Back door bees

Orchids are very picky about their pollinators. In a study of 456 orchid species, it was found that 67 percent had only one pollinating species and 14 percent limited their pollinators to two lucky species. The subfamily Cypripedioidaea, of which moccasin flower is a member, averages a less-limiting 6.3 pollinators per species. Uniquely designed flowers that trigger, trick, and trap pollinating insects are part of this species-specific behavior. Pink Ladyslippers attract small mining bees (*Andrena* species) and leaf-cutting bees (*Megachile* species) by the mock scent of nectar. Crawling into the slipperlike pouch, the bee licks or eats the hairs for the droplets of sticky fluid they exude. But now it is trapped by the slippery internal walls of the pouch and its inrolled edges and thus is ushered out the "back door" beneath the anthers. The pasty pollen packs (pollinium) are slapped onto the bee's back on its way out and held in place by a sticky glue during its foraging flight. I can attest to the quality of this glue: I once attached a pollinium to my finger and shook my hand as hard as I could, and yet it stuck solid. The next ladyslipper visited is pollinated as the bee again exits via the back door, passing under the stigma's sharp bristles that scrape the pollen pack from the bee's back and onto the stigma. Mission accomplished, the plant withers. Bumblebees (*Bombus* species), which are too large to squeeze out the back door and thus are trapped in the pouch, often eat their way out and mutilate the plant in the process.

The princess and the frog

Seeds with no nutritive packet. Roots with no rootlets. Help! How does the Pink Ladyslipper survive? The secret is in the soil, and it's a fungus. Here's how it works. Once pollinated, the ovary produces roughly six hundred thousand dustlike ovules, which mature to seeds about $\frac{1}{4}$ millimeter long, encased in a 1-millimeter-long wing. But these seeds are but embryos, virtually lacking any endosperm (the nutritive packet that fuels seed growth). Strike one. In addition, very few of these are fertile. Strike two. The last barrier to germination is the difficulty of hooking up with the right soil fungus. This is strike three for 99.999 percent of the seeds. That leaves six who hook up with the proper acid-loving endomycorrhizae (fungus roots), in this case *Rhizoctonia*. It actually penetrates the seed, forming coils that "hand deliver" carbohydrates (stored in the seeds of most plants), allowing the seed to germinate. But the job is not done for the endomycorrhizae; it lives symbiotically with the Pink Ladyslipper for the span of its life, providing the roots with food, water, minerals, and carbon during the seedling stage (unique to orchids). In exchange, *Rhizoctonia* is provided with a home inside the ladyslipper's roots. The moccasin flower could not survive without it. And they lived happily ever after.

 Sparky says: Got those midwinter blues? Yearning for a bit of color in your life? Well, why not take a mini tropical vacation? That's right, no airfare, no hotels, and probably just a short walk, bus, or drive from your house. Look for the nearest steamy greenhouse filled with exotic flowers, orchids, carnivorous plants, and maybe even a palm tree or two. Most universities have one. Larger towns might even have a public conservatory. Check it out. A stroll through a humid jungle on a gray January day will give anybody's spirit a lift.

Spotted Coralroot
Corallorhiza maculata

Clumps of reddish purple stems sprout under Northern White Cedars, Balsam Firs, Black Ash, and Quaking Aspen. Reaching to 20 plus inches in some cases (usually 8 to 12 inches), the stems have only insignificant sheathlike leaves. About twenty flowers are arranged in a loose spiral around the top half of the stem. The orchid-type flowers are $1/2$ inch across, with upper sepal and petals united to form a hood. The other two sepals wing out to the side of the lower liplike petal, which is white spotted with purple. These orchids bloom as early as mid-June and can be found in flower through much of July.

Like most orchids, the flower itself is designed specifically for efficient pollination. In Spotted Coralroot, the single stamen is fused with the style and stigma into one unit called the column. As an insect lights on the lower petal landing pad, it is guided toward the column by two parallel ridges. Working its way into the flower, it passes under the column, triggering the anther lid to open and stick two pollen masses (pollinium) to its head. But unable to get to the nectaries of these young flowers, it flies on to an older flower, where it must pass the stigma on its way to the sweet nectar. In the process the pollen is passed and fertilization is achieved. Flowers droop once fertilized, maturing into nodding brown seed capsules about $1/3$ inch long.

A true saprophyte

Spotted Coralroot is a true saprophyte, gaining nourishment from decaying organic matter and not via the sun and photosynthesis. Instead of roots, coralroots possess an underground mass of crusty branched rhizomes that entwine with the mycorrhizae of soil fungi. The fungus transfers nutrients from decayed vegetation to the plant. No photosynthesis takes place, so there is no need for green chlorophyll, and therefore the stem and vestigial leaves are reddish purple. Another clorophyll-less plant in this book, the waxy white Indian-Pipe, was also once thought to be a true saprophyte, but recent research has shown it to be an epiphyte, feeding indirectly from the roots of green plants via an intermediate fungus.

Entire plant is purple-red (occasionally yellow-brown)

Tiny orchid-type flowers have purple-spotted white lip

Grow in clumps under Northern White Cedar and Balsam Fir

8 to 20 inches tall

Blooms June 15 to July 15

 Sparky says: Subtle beauty is everywhere in the North Woods, and every individual's perception of beauty is unique. Here is an eye-opening game for the whole family. It's called camera, and here's how it's played. One person is the photographer and one the camera. The photographer leads a blindfolded or closed-eyed camera to a scene he or she feels would make a dramatic picture. After positioning the camera, the photographer opens the shutter (the camera's eyes) by pulling on his or her ear. The camera gets to take in the sight for three to five seconds before the photographer closes the camera's eyes by again pulling on an ear. Now switch roles. Use your imagination. Choose unusual angles and perspectives. Lay your camera on its back or place it on its hands and knees. Do close-ups as well as panoramics. As I said before, it's an eye-opening experience that gives you insight into the vision of others.

Wildflowers of the North Woods and Boundary Waters

Family Nymphaeacea (Water-Lilies)
- ❏ Bullhead Water-Lily   *Nuphar luteum* subsp. *variegatum*
- ❏ Fragrant White Water-Lily   *Nymphaea odorata*

Family Ranunculaceae (Buttercups)
- ❏ Marsh Marigold   *Caltha palustris*
- ❏ Red Baneberry   *Actaea rubra*
- ❏ Wild Columbine   *Aquilegia canadensis*

Family Sarraceniaceae (Pitcher-Plants)
- ❏ Pitcher-Plant   *Sarracenia purpurea*

Family Droseraceae (Sundews)
- ❏ Round-leaved Sundew   *Drosera rotundifolia*

Family Ericaceae (Heaths)
- ❏ Wintergreen   *Gaultheria procumbens*
- ❏ Creeping Snowberry   *Gaultheria hispidula*
- ❏ Small-fruited Bog Cranberry   *Vaccinium oxycoccus*

Family Pyrolaceae (Wintergreens)
- ❏ Pipsissewa   *Chimaphila umbellata*
- ❏ Pyrolas   *Pyrola* species
- ❏ One-flowered Wintergreen   *Moneses uniflora*
- ❏ Indian-Pipe   *Monotropa uniflora*

Family Primulaceae (Primroses)
- ❏ Starflower   *Trientalis borealis*

Family Rosaceae (Roses)
- ❏ Wild Strawberry   *Fragaria virginiana*
- ❏ Wild Raspberry   *Rubus strigosus*
- ❏ Thimbleberry   *Rubus parviflorus*

Family Onagraceae (Evening-Primoses)
- ❏ Fireweed   *Epilobium angustifolium*
- ❏ Common Evening-Primrose   *Oenothera biennis*

Family Cornaceae (Dogwoods)
- ❏ Bunchberry   *Cornus canadensis*

Wildflowers of the North Woods and Boundary Waters

Family Loranthaceae (Mistletoes)
- ❏ Eastern Dwarf Mistletoe   *Arceuthobium pusillum*

Family Araliaceae (Ginsengs)
- ❏ Wild Sarsaparilla   *Aralia nudicaulis*

Family Campanulaceae (Bluebells)
- ❏ Harebell   *Campanula rotundifolia*

Family Caprifoliaceae (Honeysuckles)
- ❏ Twinflower   *Linnaea borealis*

Family Compositae (Composites)
- ❏ Large-leaved Aster   *Aster macrophyllus*
- ❏ Spotted Joe-Pye Weed   *Eupatorium maculatum*

Family Alismataceae (Arrowheads)
- ❏ Broad-leaved Arrowhead   *Sagittaria latifolia*

Family Liliaceae (Lilies)
- ❏ Clintonia   *Clintonia borealis*
- ❏ Rose Twisted Stalk   *Streptopus roseus*
- ❏ False Lily-of-the-Valley   *Maianthemum canadense*

Family Iridaceae (Irises)
- ❏ Wild Iris   *Iris versicolor*

Family Orchidaceae (Orchids)
- ❏ Pink Ladyslipper   *Cypripedium acaule*
- ❏ Spotted Coralroot   *Corallorhiza maculata*

Ferns and Other Nonflowering Plants

Polypody Ferns

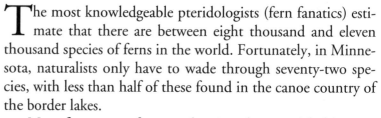

The most knowledgeable pteridologists (fern fanatics) estimate that there are between eight thousand and eleven thousand species of ferns in the world. Fortunately, in Minnesota, naturalists only have to wade through seventy-two species, with less than half of these found in the canoe country of the border lakes.

Most ferns grow from underground perennial rhizomes, which can lead to large stands of a single species. In the spring, up push the new fronds, or leaves, tightly curled like the head of a fiddle, and so the common name, fiddlehead. This growth form serves to protect the fragile fern tip during emergence. Fully unfurled leaves (called fronds) are often divided and subdivided and occasionally divided again. Henry David Thoreau theorized that "God created ferns to show what He could do with leaves." The typical lacy fern is but one form that pteridophytes can take. Check out the adder's-tongue family (Ophioglossaceae), or the water clover family (Marsileaceae), or the mosslike floating fern family (Salviniaceae), and you'll find all your fern stereotypes crushed. I somehow just can't envision a Lance-leaved Grape-Fern (*Botrychium lanceolatum*) hanging in macrame from my living room ceiling.

Considered more primitive than flowering plants, ferns reproduce via asexually produced spores. Once shed and in a suitable substrate, spores grow to form tiny leaflike gametophytes that contain an egg and sperm. Water is essential to the sperm's motility. Once fertilized, the sporophyte form of the fern grows from the gametophyte's edge. The gametophyte withers away as the fern sprouts its own roots and matures. In the Polypody family, spores are carried in round structures (sori) on the undersides of leaves.

Bracken Fern (*Pteridium aquilinum*)
aka bracken, grande fougère (Quebec)

Bracken is very common in sunny open areas, but it also likes the woods. The distinctive shape of one vertical stem terminating in

a fan of three blades is characteristic. Fronds can be tripinnate-pinnatifid, which means the leaf is divided into leaflets that are divided twice again. Brackens usually top out at waist height, but can be head-tall. Fertile fronds and sterile fronds are identical except for the spore-bearing sori along the rolled-under margins of the fertile frond. Bracken is one of the first ferns to turn brown due to early fall frosts.

This innocuous-looking fern has a hidden sinister side. Our first clue should be the fact that it contributes to no known food chain. In fact, it seems to repel insects. (Try putting a Bracken Fern upside down on your head on your next walk through buggy woods.) Horses munching a few mouthfuls keel over quickly. Autopsies reveal their vitamin B reserves to be destroyed. Cattle suffer a similar, but slower, fate. In something akin to radiation sickness, or cancer, their bone marrow quits functioning; as a result, platelets are not produced, so blood cannot clot. The cows, literally weeping tears of blood, die from internal bleeding. If this isn't bad enough, recent medical findings show that the delicious spring fiddleheads (sauteed in butter . . . mmmm!) are carcinogenic!

How about this for a mystery? A fellow naturalist friend of mine has noticed the brackens of Polly Lake in the Boundary Waters Canoe Area Wilderness to be riddled with insect-chewed holes. Now remember, nothing is supposed to eat this fern. He's visited the island site during every daylight hour from dawn to dusk on several occasions, but as of yet has seen no sign of any minuscule munchers. This is one for "Unsolved Mysteries."

Rock Polypody (*Polypodium virginianum*)

This small (4 to 10 inches) evergreen fern can be found clinging to cliff faces, hiding in rock crevices, and growing on mossy boulder tops, even peeking out from under the snow in midwinter. The fern consists of a single leathery blade growing from a rhizome, which is pockmarked with the scars of old leaves snapped cleanly off. Leaflets are rounded, smooth-edged,

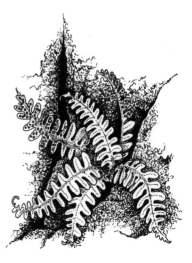

and not cut all the way to the stem. Perfectly round rust-colored sori dot the undersides of leaves from July to September.

Sensitive Fern (*Onoclea sensibilis*)

"Sensitive to what?" I ask. Does it curl up in response to human touch? No. The best (i.e., only) answer I've heard is that it is sensitive to early frosts, withering quickly thereafter. It also shrivels up rapidly after being cut. Look for the Sensitive Fern in wet spots. Fifteen to 30 inches tall, the lone blade is deeply lobed but not cut all the way to the stem. Leaflet tips are blunt. A lacy network of veins covers the blade. The ball-like sporangia are carried on a totally separate fertile stalk (rare in the Polypodiaceae family), which is much shorter than the fronds and brown when ripe. It is very noticeable in fall and early winter. Spores may be produced on the stalk for two or more years. You can distinguish these dried fertile fronds from those of Ostrich Fern by the fewer branches (only six or seven) and very definite ball-shaped spore cases of the Sensitive Fern.

Ostrich Fern (*Matteuccia struthiopteris*)

The king of ferns when it comes to height, the Ostrich can reach 8 feet, though 3 to 6 is average. The fern is named for the frond's resemblance to the plume of an Ostrich, getting wider above the center and terminating rather abruptly. Clumps originate at a common point and arc away from the center. In the middle arises the shorter fertile stalk. Brown when ripe, the fertile stalk bears many branches of tubelike cases that hold the spores. It, like the sterile fronds, is plume-shaped and will last through the winter. Colonies are found in the rich, wet soils of stream and river floodplains and in damp woods.

In the spring, rootstalks send up the tightly coiled new leaves. Resembling the head of a violin, they are known as fiddleheads. This growth form results from faster growth on the lower than upper surface of the leaf. The fiddleheads of Ostrich Fern are encased in a woolly cover, which detracts not at all from their delicate taste when sauteed in butter and salted. In fact, this is the same fiddlehead that is served in the finest gourmet restaurants.

Sparky says: Do you have a shady spot around the house where nothing seems to grow? Maybe that is the perfect spot for a fern garden. The Ostrich Fern is probably the most common domesticated yard fern because of its hardiness and tolerance, but there are others out there to try.

You could also try your hand (or should I say your green thumb) at shade gardening with wildflowers. More and more wildflower seed companies are springing up, and most have special mixes for deeply shaded areas.

Flowering Ferns

Osmunda species

Fronds arch to more than 2 feet

Stem is a massive rhizome whose top forms a mound above ground

Spores are never found on the undersides of leaves

Inhabit wet areas

Other names

   Royal fern family

The "flowering" ferns are a small family of large and conspicuous true ferns that, of course, do not flower. They are made conspicuous by their large size and unique spore clusters. Most sporangia of most ferns are located directly on the undersides of a fern's leaves, but not in the family Osmundaceae. Theirs are borne in clusters on stalks. The clusters are brownish when ripe. Spores are short-lived, though, having only a two-week window in which to germinate. Flowering ferns sprout from perennial, underground stems, or rhizomes, which bulge to the surface and form noticeable mounds at the base of the fronds. These root masses, known in the nursery business as osmunda, are used to cultivate epiphytic orchids and other plants that do not require soil.

Cinnamon Fern (*Osmunda cinnamomea*)

These ferns form clumps of 2- to 3-foot long fronds that arch backwards. In the center rise several totally fertile stalks bearing thousands of cinnamon-colored spore cases. They are nearly as tall as the fronds, but wilt once the spores are released in early summer. Leaves are very similar to those of Interrupted Fern but are more pointed at tips and hairy along the margins. Cinnamon Fern grows in bogs, marshes, and swamps, preferring a wetter site than the Interrupted Fern and a drier site than the Royal Fern.

How they arrived at this belief is beyond me, but the pioneers thought that one bite of an unfolding frond prevented the agony of toothache for a full year.

Interrupted Fern (*O. claytoniana*)

Here is yet another descriptive and extremely helpful fern name. Interrupted Fern fronds are indeed interrupted by one to five pairs of short stalks laden with green to brown round spore cases. They are located about two-thirds of the way up the frond and are evident at a glance. Even after the spores are dispersed and the cases disintegrate, the frond is still left with a noticeable gap. Fronds, which may reach 5 feet, have blunt-tipped pinnules (leaflets). This species is found in moist and shady woods, preferring the driest soils of any of the "Three Marsh-keteers."

Royal Fern (*O. regalis*)

Fiddleheads of this species emerge wine-colored. They open to reveal very unfernlike fronds. The leaflets are rounded and smooth-edged, not divided and subdivided as in many lacy species. They almost resemble the compound leaves of some trees. Spore cases are borne in a dense head at the frond's tip. Royal Ferns can reach 5 feet tall, but usually grow only 2 to 4 feet. Common along lakeshores in the canoe country, Royal Fern can also handle the acidic waters of marshes, swamps, and wet woods. It can tolerate shade or full sun. Yellowing in autumn, Royal Ferns become quite noticeable to the fall paddler.

The Royal Fern had an incredible reputation with early settlers. They believed it possessed the power to heal wounds and mend broken bones. Some even claimed eternal life was yours if you drank of the sap.

Osmunder the Waterman

Family Osmundaceae is named for the Saxon god Osmunder the Waterman, who hid his family from danger in a clump of these ferns.

Sparky says: Make your own autumn greeting cards. You'll need a box of blank cards and envelopes, which are readily available at any print shop (4" x 5" is a nice size), high quality clear contact paper, scissors or Exacto knife, and a flair for two-dimensional flower arranging (optional!). Now the fun part. On your next fall foray, gather some interesting foliage (yellowed fern fronds, crimson maple leaves, feathery seeds, evergreen leaves). Just remember, the flatter these items will lie, the better your cards will turn out. Arrange your autumnal treasures on the front of the card. Peel the backing off the contact paper and place a piece slightly larger than the card over it, being careful not to trap any air bubbles or make any wrinkles. Smooth firmly with your thumb. Trim excess contact paper. Voila! You are an instant artist. Since the colors will fade, it's best to write your friends ASAP. When's the last time they heard from you, anyway?

Clubmosses
Lycopodium species

Common ground cover in the boreal forest

Stems creep over ground

Short, stiff evergreen plants

Form single-species colonies

Spores often borne in slender, erect "cones" atop long stalks

Other names

    Lycopodium

    snake moss

    running moss

    lummer (Swedish)

The clubmosses are not mosses. Rather, they are nonflowering vascular plants that are considered, along with horsetails, to be intermediate between higher plants (trees and flowers) and those with no conductive tissue (lichens, mosses, liverworts, and fungi). True mosses do not possess xylem or phloem and so are nonvascular, and hence considered more primitive than clubmosses. Members of the *Lycopodiaceae* grow along branching horizontal stems that crisscross the ground, sending down roots as they grow. Colonies can cover large areas of forest floor. Old stems wither, but the new growth is quite rapid (several inches each year) so it spreads quickly. Spores are carried in slender erect "cones," or strobiles, that are either sessile, arising directly from the stem tip, or stalked. The cones are the "clubs" that give this family their confusing common name. Each strobile gives off millions of dustlike spores in the fall. A landing insect or puff of wind is all it takes to scatter the seed. It takes roughly twenty years for a spore to grow into a spore-bearing adult plant. The upright evergreen portions of the plants are varied, and identification will be covered under each species.

Steamy jungles and giant ferns

The distant ancestors of these diminutive clubmosses were veritable giants in the steamy jungles of the Carboniserous Period, 350 and 300 million years ago. Ferns stood 40 feet tall, but the clubmosses, some 5 feet in diameter, towered to 120 feet. It is thought that the buried carcasses of these giant forests have become the coal deposits of today.

Ground-Cedar (*Lycopodium complanatum*)
aka staghorn pine

This clubmoss really does resemble a tiny cedar tree. Stem branches are composed of scalelike leaves, flattened and fanned out in one plane. The horizontal stem is above ground. One to four cones (2½ inches tall) are borne, like candelabrum arms, on 3-inch stems.

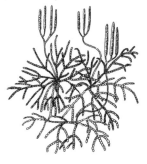

Round-branched Ground-Pine (*L. dendroideum*)

Appearing to be a seedling spruce, this clubmoss requires a closer look. Note the nonwoody stem and the 2^1/2-inch stemless strobile tipping the plant. Once you've identified one, you'll never mistake it for a tree again. Horizontal stems are underground.

Stiff Clubmoss (*L. annotinum*)
aka bristly clubmoss

Probably the most common canoe country clubmoss (say that ten times, fast!). Its military straightness and sharply pointed leaves held out at a ninety-degree angle to the stem give it a neat symmetrical look. Straw-colored strobiles 1^1/2 inches long are sessile, emerging directly from the tips of stems. Plants may reach 12 inches tall. It enjoys acid soils.

Shining Clubmoss (*L. lucidulum*)

Bright green stems reach 10 inches and may fork three times. This clubmoss is unique in not bearing spores in strobiles or cones but in tiny sporangia in the axils of the leaves. Also note that leaves are of unequal size and toothed at the tips. Shining Clubmoss prefers less acidic soils and clay pockets.

Running Clubmoss (*L. clavatum*)
aka running pine, common clubmoss, wolf's claw

Long, forking horizontal stems create a maze atop the ground. Straw-colored strobiles usually occur in pairs atop a single slender stalk of 6 inches or less. Leaf tips end in threadlike hair. This clubmoss is more common in less acid soils under aspen, fir, and cedar.

Special spores

Microphotographs of clubmoss spores reveal an incredible uniformity in size. In the early days of science, a spore was used as a standard unit of microscopic measurement. But that's not their only use. When dry, the spores can be lit to create an explosion. The flashes for early photographs, fireworks, and theaters were all made from clubmoss spores. Their water-repellant nature led to their use as coatings that kept pills from sticking together and as a talclike smoothing powder. Eczema and chafed baby bottoms were ideal targets for the magic powder. The spores were marketed as "vegetable sulphur." Blackfoot and Potawatomi inhaled the dust to stop a bleeding nose.

Clubmoss cure

Did you know that *Lycopodium clavatum* tablets temporarily cure minor aches and pains associated with a sore throat, specifically one that is soothed by warm drinks, is worse on the right side, and hurts more from 4 to 8 P.M.? I didn't either until I saw them advertised in this way at the Whole Foods Co-op.

Sparky says: Get on your belly in a patch of clubmoss. Look into the Lilliputian forest. Try to imagine the small stems as the 120-foot-tall plants their extinct cousins once were. Watch out for that giant dragonfly with a 3-foot wingspan!

Part two: Now for you adult kids. In late summer and late fall, when the clubmoss strobiles are shedding spores, try lighting a head. Historically used as the flash powder for photography, the spores should ignite. Hold your lighter near the spore-laden cone of a clubmoss, flick your Bic, and then flick a head. Poof! Hold on to your eybrows!

Christmas wreaths

While driving up the North Shore of Lake Superior, listening to radio station WIMI from Ironwood, Michigan (the signal skips well across the lake), I heard a holiday wreath-maker's plea for "ground pine." They paid by the pound. The ripped-up clubmoss was made into long-lasting Christmas wreaths. But I wondered, can we afford wholesale destruction of this unique plant for a seasonal throwaway item?

Horsetails

Equisetum species

Plant of the moist and wet areas

Hollow stem is stiff, ridged, and conspicuously jointed, with pop-apart sections

Fertile stalks have a whorl of scalelike leaves at each node and a strobile on top

Vegetative stalks have a feathery appearance due to whorls of branches at each node

Other names

    scouring rush

    Equisetum

    Indian Tinker Toys

    fraken (Swedish)

Indian Tinker Toys is what we called them as kids. Pulling apart the fertile stems at each node and trying to snap them back together was about as much fun as a kid could have—for ten minutes, anyway. What we didn't know then was that horsetails, along with clubmosses, are the links between flowering plants and nonvascular plants. They do contain xylem and phloem, which allow transport of food and water from the roots, but bear spores instead of seeds.

Colonies spring up from a perennial underground rhizome network every year. Most species like moist soil, some even thriving in shallow water. Fertile stems are stiff but hollow, with a distinctly ridged surface. Embedded silica in the epidermal cells makes them rough enough to file fingernails and scrub pots. Horsetails have noticeable nodes ringed by scalelike leaves that hug the stem. Spore-dispersing strobiles, or cones, top the fertile stalks. The vegetative shoots are much finer and appear feathery due to the long wispy branches whorled at each node. It is from this form that the name horsetail was derived. Some species bear strobiles on this type of shoot. Spores landing on unsuitably dry or rocky ground uncoil four narrow wings, or elaters, that presumably allow the wind to carry them further down the trail, so to speak, and on to greener pastures where it is wet enough to make germination possible.

Wood-Horsetail (*Equisetum sylvaticum*)
aka forest horsetail

Fertile stems at nodes have characteristic reddish brown "teeth," or scalelike leaves, which are clumped together in threes. Stems bear sawlike spicules of silica on the ridges. When young, the fertile shoots are brown, but they quickly sprout green branches. Once the spores are released and the cone withers, the fertile stem will resemble the vegetative, or sterile, plant, complete with long, double-branched branches. Fertile stems grow to 12 inches, sterile stems to 20 inches. This is the most common canoe country woodland horsetail.

Field-Horsetail (*E. arvense*)

This is the only northern horsetail whose fertile stalk lacks chlorophyll, remaining brownish until the cone withers and dies. (Wood-Horsetail, above, is brownish at first, but quickly sprouts green branches.) The sterile Field-Horsetail has long, green branches angling up from the stem. It may reach 14 inches. Growing in open, sunny, disturbed sites, it can serve as an excellent soil binder, preventing further erosion. Watch for it along river banks.

Water-Horsetail (*E. fluviatile*)

Most common in very wet sites: bogs, swamps, and water from three inches to three feet deep. Water-Horsetail takes only one form. The fertile stem can be quite tall for having such thin walls and a large hollow center. Fifteen to 20 or more teeth encircle the nodes. Branches, if present, are sparse. It is very common in lakes and rivers of the North Woods.

Herbal healing

Since silica enhances the body's ability to absorb and use calcium, a horsetail tea has been prescribed for those with white spotting on fingernails, cracked nails, and lifeless hair. Silica has also been used to treat people whose lungs have been damaged by tuberculosis. Since it is astringent, it helps stop bleeding in stomach ulcers and has recently been shown to curb bed-wetting in children. Horsetail is a specific for treating inflamed and enlarged prostates. The high silicic acid content also nourishes connective tissue, and so it is included in remedies for rheumatism and inflamed tendons. Because it is a strong diuretic (increases the flow of urine), it should never be used during pregnancy.

Calamite calamity

Giant horsetails known as calamites lived during the Carboniferous Age, growing side by side with the massive ancestors of

Scouring rush

Pioneers used horsetail stems (especially *E. hyemale*) to scour the crud off their metal pans, and hence the common name "scouring rush." It is the silica deposits along the stem that make it such an abrasive plant. Tiny teeth, or silica spicules, make certain species' stems' vertical ridges look like a saw blade under magnification. Silica is the stuff that quartz (chert, flint, sand) is made of and is also found in sponge skeletons and shells. Try scrubbing your pots or pewter clean sometime, but always wash them out thoroughly before cooking in them.

clubmosses and ferns. What a strange sight to see a horsetail 60 feet tall and a foot in diameter! And its form was nearly identical to today's diminutive descendants. After burial, these strange trees were converted to peat by pressure and eventually became either hard or soft coal, petroleum, or natural gas. The gas that you drove to work with today may have been at one time a giant horsetail of the Coal Age. Think about that for a minute.

Sparky says: Fingernails a little ragged after a week in the woods? Try horsetails! Pluck a single fat fertile stalk. Fold it several times and go at it. Rasp those ratty edges to a hygienic smoothness. We call this the "Minnesota manicure."

While I'm on the subject of personal hygiene, how's your hair? Lifeless and tangled, you say? Well, you need the herbalist's horsetail hair rinse and tonic. Gather about eight stout stems of the fertile variety, and pluck them apart node by node. Place in a pan and mash them up a bit. Pour a pint of boiling water over them and continue mashing. Cover and leave until lukewarm, at which time you strain off the liquid. Quickly shampoo and rinse your hair. Now, work in the horsetail infusion. Massage it into your hair and scalp. Blot up excess with a towel, but do not rinse! Comb through your hair and dry in your normal fashion.

This tonic should "help restore the scalp's natural acidity and strengthen the circulation giving a healthy shine to hair." Its saponins create the lather, flavone glycosides stimulate circulation in the scalp's tiny blood vessels, and the silica gives your hair body.

The shag carpeting that covers the Canadian Shield is a mosaic of sphagnum moss species. Some species are specialists on the mineral-rich sedge mats where the pH is a neutral 7 (*Sphagnum teres* and *S. subsecundum*). Others thrive in the acidic sphagnum "lawns" ringing bogs. *Sphagnum fuscum*, a circumpolar species, occupies the most acidic niches, spots where the pH may be as low as 4. In the spruce muskegs, more shade-tolerant mosses thrive. The abundant *S. magellanicum* (Red Spoonleaf Peat Moss) lives here. With its reddish fat leaves and rosette head, it is easy to identify. In the shade it may be more pink to green than red, but it always forms extremely spongy round-topped hummocks. Individual stems may be 8 inches tall. It is dominant in spruce bogs along with *S. recurvum*. Northern White Cedar swamps are a unique habitat with soil that is high in calcium. Therefore, mosses must be able to tolerate a higher pH soil and deep shade.

Sphagnum mosses have five branches at each node along the stem, more dense at the apex, forming a tufted head. Leaves lack midribs. They are usually erect since cells are filled with water. The presence of anthocyanin pigments results in their often taking shades of red and purple.

Spore cases are spherical and on short stalks. As the internal tissues of the case shrink, gases build and internal pressure increases. Finally, the case explodes, blowing the top off and shooting up a tiny cloud of spores.

More absorbent than Bounty paper towels

Sphagnum moss is able to absorb and hold up to twenty-seven times its weight in water. Let's see Bounty paper towels top that! This is three to four times more than cotton. Water is merely enmeshed in the cotton fibers, but sphagnum leaves actually have a spongelike structure that sucks up and traps the water. Large dead cells riddled with pores serve as water holding tanks. These cells contain no chlorophyll. Balloon-shaped and reinforced with spiral bands, they are designed, instead, for absorption.

Sphagnum Mosses
Sphagnum species

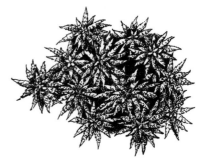

Stout, thick-leaved mosses

Red to pink to green in color

Stems may reach 8 inches long

Found in floating bogs, spruce muskegs, sedge mats, and cedar swamps

Other names

    asakumig (Ojibwa)

    peat moss

    vitmossa ("white moss" in Swedish)

Pollution indicator

Walter Glooshenko, research scientist with the National Water Research Institute in Burlington, Ontario, has pioneered the use of mosses as air pollution indicators. By testing the circumpolar moss, S. fuscum, Glooshenko can compare pollution levels in Canada, the U.S., Scandinavia, and Siberia.

This sponging ability has led to numerous uses through time. The Ojibwa made diapers out of the stuff. Free, absorbent, sterile, disposable, and biodegradable, what could be better? After drying the sphagnum over a fire, they mixed it with rotted cedar and a "wool found in seed vessels of a species of reed." This was placed in a birchbark tray and placed under the baby in its cradleboard. To change diapers, one simply flung the contents of the tray into the woods.

Blood has been mopped up with sphagnum for centuries. Medical staff during the Napoleonic Wars of the eighteenth century and the Russo-Japanese War of the 1880s employed it as surgical dressing. But it was not until the Allied Forces ran out of cotton bandages in World War I that moss pads were mass produced. The precious and dwindling stores of cotton were needed for gun wadding. By 1918, the British were using one million sphagnum pads each month. Scotland and Canada were the major manufacturers of the cotton-gauze-covered, moss-filled bandages, cranking out up to four million a month. They were cheaper to make, cooler, less irritating, and absorbed three times the liquid three times faster than the all-cotton pads. And wonder of wonders, wounds treated with sphagnum pads seemed to heal faster. Today we know that microorganisms associated with sphagnum moss fight infection with their antibiotic activity, while small quantities of natural steroids speed healing. This revelation would have been nothing new to the Alaskan Indians, who've used a salve of sphagnum leaves and grease to treat cuts for centuries. Many European and Indian groups have also used the moss as menstrual padding.

Making its bed

Sphagnum moss alters its environment, even to the point of making the habitat unsuitable for its own existence. First of all, it is able to exclude other plants by increasing the acidity of the water. It does this by soaking up rain that is already acidic, binding calcium, magnesium, and the like in cells that contain organic acids, and releasing hydrogen ions, which lower the pH of the water. Not much, but sphagnum can survive in the vinegarlike water.

Sphagnum forms the floating mat that rings most bogs. Growing from the outside perimeter towards the center, the moss lawn will eventually cover all the open water. Walking on such a bog is like walking on a trampoline whose skin is too loose. Gradually, sinking moss and roots of invading Leatherleaf fill in the space. When it's dry enough and solid enough, trees such as Tamarack and Black Spruce encroach, turning the onetime bog into muskeg. The water-loving sphagnum species are gone, but they are replaced by other species.

Over time, many generations of sphagnum die and are packed into the anaerobic dregs of the bog or muskeg. Here the moss does not decay but is compressed over hundreds of years into peat. Yes, the stuff Irish diggers cut from the ground in nice blocks to burn at home for heat. Minnesota has vast quantities of peat, but commercial harvest and use is, for the moment, limited.

Sparky says: Test for yourself the amazing absorbency of sphagnum moss. Pick a double handful of the deepest, spongiest moss you can find. Get it from an out-of-the-way place that is not near a trail or campsite. Dry it in the sun or over a fire. Now put a layer of water in a pan and try to soak it up with your moss. Wring it out. Compare the weight of similar-sized bundles of dry and soaked moss. Do you agree that sphagnums can absorb twenty times their weight in water?

Miscellaneous moss

Since there's so much of it around in the North, people have come up with all kinds of uses for sphagnum. Stuffed between logs of a cabin, moss chinking keeps out the cold winter winds . . . kind of. Made into mukluk liners by the Inuit, it keeps the warmth in. In China it has been a starvation food for humans. The Sammi of Lappland have upgraded it to a staple by using it in bread. Soaked in tar, it was used to caulk boats. Soaked in molasses, it was fed to livestock. You can make a brown dye from it. Burning the stuff creates a smoky fire, which fends off mosquitoes and keeps away cold in fields of crops threatened with freezing. It can be woven with wool, mixed with clay to make lightweight bricks, or blended into cement to make "peatcrete."

Lichens

Freddie Fungus and Alice Algae took a "lichen" to each other and now live together in "sym-biosis," but their union is "on the rocks" or so the old tale goes. Indeed, lichens are actually two organisms joined into one plant. The algae are free-living green plants that produce their own food by photosynthesis. They can survive without the fungus, but not vice versa. The fungus has a rootlike system (mycelium) that gathers mineral nutrients from substrates (rock, wood, soil) and holds water. Once they hook up, the fungus seems to take over, providing the basic shape of the lichen as the cell walls thicken. It provides shelter and salts for the algae in return for food, but the algae is clearly held in check. Mycologists now believe that this partnership is not one of symbiosis, but rather a mildly parasitic one in which the fungus dominates.

Lichens inhabit nearly every corner of this big blue ball we call earth. Arid desert to Arctic tundra, bare rock to ocean floor, and everywhere in between, we find the 15,000 to 20,000 species of lichens that inhabit the world. North America has 2,800 species and Minnesota 600. Yet, unlike flowering plants, which increase in abundance and variety as you near the equator, lichens decrease—they become more common towards the poles. Antarctica lists 350 species but only two vascular plants. Seven lichen species even eke out a life in the Queen Maud Mountains, which sit at 86° 83° south latitude, nearly at the bottom of the world! Lichens are survival experts.

It was once thought that lichens were successful because the fungal portion protected the algal part from drying out. Today, it is believed that rapid desiccation is their key to survival in extreme environments. When lichens dry out to between 2 and 10 percent water content, the outer walls thicken and become more opaque, shutting out sunlight. Photosynthesis shuts down, and the plant is able to tolerate extreme cold and scorching heat. This temporary slowdown happens, not only with the changing seasons, to a lesser degree, in the course of a day. Morning dews moisten the lichen, and photosynthesis kicks in until the sun dries them out. Ideal photosynthetic rates are obtained when the lichen is at 65 to 90 percent of total water saturation.

Old timers

Lichens grow slowly . . . very slowly. In the North, crustose lichens on rock may expand their diameter by less than one millimeter per year. Using this measurement, lichenologists have determined that some lichens in the Arctic may be about 4,500 years old. This would make them, along with aspen clones and Bristlecone Pines, candidates for the world's oldest living organisms.

Pollution police

Lichens make excellent biomonitors since they absorb all their nutrients from the air. In soaking up a recent rainfall, they metabolize useful nutrients, but pollutants, such as lead, mercury, zinc, and sulfates, permanently bond to the lichen's cell walls in concentrations proportionate to those in the atmosphere. Mason Hale, lichenologist at the Smithsonian Institution, has found sulfates in lichen populations deep in the pristine wilderness of Wyoming's Wind River Range, the direct result of acid rain. Sulfur dioxide is released into the atmosphere by burning fossil fuels, such as coal and gasoline, and falls to earth as acid rain. Not a picky eater, the lichens absorb any and all airborne compounds. The National Park Service is now monitoring air pollution via lichens and fancy monitors.

April 28, 1986, was the day of the great Chernobyl nuclear disaster, and its effects are still being felt today. The radioactive fallout was heaviest in central Sweden and Norway. Strontium 90 and cesium 137 were readily absorbed by reindeer lichens (*Cladina* species) and in turn, eaten by the Sammi's reindeer herds. When it came time to butcher reindeer, tests showed the meat of some Swedish animals had radiation levels of 62,200 units per pound. Acceptable levels were around 680 units per pound. Millions of pounds of useless reindeer

Rock crusher

Lowly lichens are the pioneers of bare rock in many areas. They gain a foothold by actually dissolving rock minerals with their lichenase acids. After a minute crack forms, water can get in, freeze, and break it apart even more. Dust and dirt blow in. Once there is a 10 percent build up of organics, mosses can grow. Ferns, flowers, and firs follow.

Dyed in the wool

Lichens were used exclusively in the dyeing of the famed Harris tweeds. Acids from the plant even made the wool resistant to attack by clothes moths.

meat was dyed blue and buried, a crushing blow to the Sammi of Lappland, whose livelihood depends on that animal. And not only reindeer, but berries, sheep, and cow's milk were all deleteriously affected. Since the half-life of strontium is twenty-eight years and cesium thirty years, the effects of Chernobyl are still being felt and will be for a long time to come.

British Soldiers (*Cladonia cristatella*)
aka British redcoats, scarlet-crested cladonia

Usually tiny plants, especially nonflowering types, get little recognition from the general public. But anyone who spends any time camping and canoeing knows this lichen with the knobby red tops. It is called British Soldiers after the fighting men of the royal army and their scarlet battle tunics. Though only 1/2-inch tall, this lichen garners "oohs and aahs" when pointed out. It grows in the company of Pixie Cup and ladder lichens on dead wood and on the thin soil of bare sun-splashed granite. It is at its peak in spring, when the stalks are light green and the fruiting heads a shining scarlet.

Pixie Cups (*C. pyxidata*)

These 1/2-inch-high gray goblet-shaped lichens are where the pixies, those cheerful yet mischievous sprites, take their baths . . . or so I've heard. Actually, the open top of the plugged funnel is where the sporelike soredia (fungal hyphae entwined about a colony of algal cells) rest. When a raindrop splashes into the cup, the soredia are sent flying, but a new lichen will grow only if it hooks up with its host algae, in this case *Pleurococcus*. Various-sized Pixie Cups grow together in groups on dead wood and soil-covered rocks. They like sunny places.

Old Man's Beards (*Usnea* species)
aka beard lichen, Usnea, hairlike Usnea

The greenish old man's beard hangs in tufts from the branches of trees, looking very beardlike. Many mistake this North Woods lichen for a moss, especially Spanish Moss (*Tillandsia usneoides*), which drapes the branches of southern trees. Spanish Moss is actually the threadlike leaves and stems of a flowering plant and, like *Usnea*, not a moss at all. Dead Balsam Firs festooned with *Usnea* appear, at a distance, to be quite alive. It can also grow on live conifers. Fibers are 4 to 10 inches long and flexible. If you cut one in half under a stereoscope, you'll see that the core is a tough fiber surrounded by spongy stuff. Fruiting bodies are lighter-colored disks edged with fine hairs.

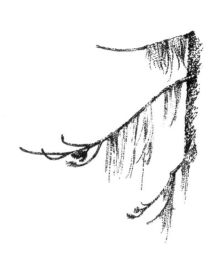

The doctrine of signatures was a medieval belief that God shaped plants like the things they cured. It is no surprise that *Usnea* was used to treat disorders of the scalp. Even earlier than this, though, Theophrastus in 300 B.C. credited the lichen with the power to grow hair. Sure—maybe if you glue enough of it to your bald head!

The tiny boreal warbler called the Northern Parula constructs its cup nest of grass deep in the recesses of a dense tuft of old man's beard, preferably high up in a spruce or fir. Olive-sided Flycatchers line their nests with the soft lichen.

Rock Tripes (*Umbilicaria* species)
aka tripe de roche, toadskin lichen, iwatake (in Japan)

This gray to black leaflike lichen clings to rock faces and woodland boulders. Rock tripe are nickel-sized to platter-sized sheets, stuck to rock by a thick cord of tissue (which is where the genus name *Umbilicaria* comes from). When dry, the crusty cracked edges upturn, revealing the velvety black undersides. After a rain, the water-soaked lichen is soft and rubbery. Some have called it a "living hygrometer."

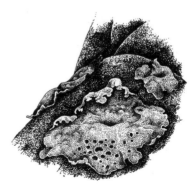

Rock tripe has a completely undeserved reputation as an edible lichen. Oh sure, Sir John Franklin and his men were saved by it on their 1823 Arctic expedition, but they also suffered extreme bowel complaints and nausea. Bitter, rock-dissolving acids were probably to blame. A little wood ash or soda added to it might have neutralized the lichen a bit. The great French explorer Pierre Esprit Radisson related this rock tripe tale from his third journey, 1658 to 1660: "The kittle was full with the scraping of the rocks, which soone after it boyled became like starch, black and clammie and easily to be swallowed. I think if any bird had lighted upon the excrements of the said stuff, they had stuck to it as if it weare glue." These guys should have taken a lesson from the Native Americans, who never touch the stuff. The Japanese, of course, consider "iwatake" a delicacy and put it in salads or deep fry it. They also do backflips over raw fish.

Reindeer Lichens (*Cladina* species)

aka reindeer moss, caribou lichen, caribou moss

Reindeer lichen forms large patches on the thin soil covering bedrock, especially in old growth Black Spruce–Jack Pine forests. Recent research in Quebec shows that *Cladina mitis* dominates in forests under seventy years old, while *C. stellaris* prefers living under stands over one hundred years old. The many-branched coral-like lichen forms round-topped bunches of varying heights and shades of gray, yellow, and green. *C. rangiferina* tends toward the grays and whites, while *C. stellaris* is more yellowish and green. Dry clusters almost hurt to the touch, they're so sharp, and walking through a patch would turn it into dust. But just a little moisture renders them downright cushy.

If you've ever seen an elaborate model train setup, you've probably seen reindeer lichen. Its round-topped, branching structure makes it an ideal double for miniature trees and shrubs.

This lichen is named after reindeer and their wild cousins, the caribou, for good reason. It is the primary winter food of these species. But since the lichen is low in sodium, the hoofed ones become salt-starved and will kill lemmings, lap up dog or human urine, and even drink saltwater just to fulfill the craving. Though they absorb only 28 percent of the lichen's carbohydrates, the caribou would become sick if they didn't eat it. Evidently, it's necessary for proper digestion.

As human food, reindeer lichen may be more palatable after partial digestion in the grazing ungulate's gut. The Dene people of the North American Arctic made a delicacy of it. After killing a caribou, they mixed some of the animal's blood with the partially digested lichens directly in the stomach, and hung it up in the heat and smoke for several days to allow for fermentation. This concoction was then boiled and mixed with fat and flesh strips. Samuel Hearne wrote in 1795 that the native dish had "such an agreeable acid taste, that were it not for prejudice, it might be eaten by those who have the nicest palates." Well, the next time I'm in the Arctic . . .

Sparky says: Here's an eye-opening activity that will utilize your child's vivid imagination. Its called "Nine-square-foot National Park" and the idea is to create a minipark full of fantastic sites. Mom, dad, or leader should play too. Each person gets a 12-foot string (this is the perimeter of 9 square feet), which can be lain anywhere, just so the ends touch. This now becomes the boundary of your minipark. Select seven interesting minisites within your borders. Connect them with a yarn or string "trail." Name your park. Name your sites. Now everyone gets to play Interpretive Ranger and give everyone else a tour of their park. Moss, fern, and lichen-covered granite bedrock works great for this activity.

Ferns and Other Nonflowering Plants of the North Woods and Boundary Waters

Ferns
Family Polypodiaceae (Polypodys)
- ❏ Bracken Fern — *Pteridium aquilinum*
- ❏ Rock Polypody — *Polypodium virginianum*
- ❏ Sensitive Fern — *Onoclea sensibilis*
- ❏ Ostrich Fern — *Matteuccia struthiopteris*

Family Osmundaceae (Flowering Ferns)
- ❏ Cinnamon Fern — *Osmunda cinnamomea*
- ❏ Interrupted Fern — *O. claytoniana*
- ❏ Royal Fern — *O. regalis*

Clubmosses
Family Lycopodiaceae (Clubmosses)
- ❏ Ground-Cedar — *Lycopodium complanatum*
- ❏ Round-branched Ground-Pine — *L. dendroideum*
- ❏ Stiff Clubmoss — *L. annotinum*
- ❏ Shining Clubmoss — *L. lucidulum*
- ❏ Running Clubmoss — *L. clavatum*

Horsetails
Family Equisetaceae (Horsetails)
- ❏ Wood-Horsetail — *Equisetum sylvaticum*
- ❏ Field-Horsetail — *E. arvense*
- ❏ Water-Horsetail — *E. fluviatile*

(Sphagnum Mosses)
- ❏ Sphagnum Mosses — *Sphagnum* species

Lichens
Family Cladoniaceae
- ❏ British Soldiers — *Cladonia cristatella*
- ❏ Pixie Cups — *C. pyxidata*

Family Usnaceae
- ❏ Old Man's Beards — *Usnea* species

Family Gyrophoraceae
- ❏ Rock Tripes — *Umbilicaria* species
- ❏ Reindeer Lichens — *Cladina* species

Fungi

Fungi

Friends, we have fungus "amung-us," and it may be as near as the itchy crevice between your little toes. Yes, athlete's foot is a fungal problem that many a wet-foot camper has had to contend with. The family of fungi also includes such diverse members as ringworm, blue cheese mold, White Pine Blister Rust, histoplasmosis, morels, puffballs, penicillin, and yeast. We are going to concentrate here on the high profile basidiomycetes, or club fungi. These include the mushrooms, shelf fungi, puffballs, and stinkhorns—species that one might encounter on a hike, paddle, or campout.

Fungi are not plants. They perform no photosynthesis to make their own food. Instead they are saprophytes, absorbing nutrients in solution from decomposing organic matter. How do they do it? The above-ground fungus is simply the spore-dispersing apparatus. It is just the tip of the iceberg, so to speak, for within the tree or soil is a mass of white threadlike hyphae (fungal roots) that secrete enzymes to break down the compounds into soluble substances. The hyphae directly absorb this food, nourishing the entire fungus.

Now we know that green, or once-green, plants feed the fungi. In return, the fungi break down dead and dying plants into simpler chemical compounds, which then can be absorbed by other green plants. Mold fungi act first, breaking down sugars and starches. Then the club fungi take over, working over the cellulose into simple chemical compounds such as carbon, nitrogen, potassium, and phosphorus. Returned to the soil, these elements become available for uptake by a new generation of green plants. Can you imagine a 5,000-year-old forest with death but no decay? It would be a mile-high jumble of trees with all the carbon compounds locked up. Fungi are major recyclers.

Amanita family—Amanitaceae

The amanitas are famous—or should I say, infamous. Responsible for 90 percent of deaths from eating mushrooms, this poisonous and hallucinogenic family is only "esteemed by maggots and mystics." But from the visual standpoint they can be esteemed by all. Large (up to 2 feet across!) and showy (red or yellow caps often speckled with whitish chips), they leap out at even the most oblivious woodland wanderer. They break through the ground as an "egg," which is the developing button-mushroom encased in a skin known as the universal veil. As the button amanita expands, the veil tears, leaving bits clinging to the cap. These veil remnants are the "butter brickle chips" that speckle many amanita caps. An annulus, a ring of skin circling the stem, is also a remnant of the veil and an identifying characteristic to look for. Lifting the mushroom gently from the ground should reveal the bulbous base called the volva.

There are exquisitely edible amanitas. *Amanita caesarea*, for example, was the favorite of Roman rulers. But this family also contains one of the most deadly mushrooms on the face of the earth, *Amanita phalloides*, the Death Cap. Just two ounces is enough to kill any human. It is alleged that this was the fatal fungus slipped to Emperor Claudius.

One last tidbit about these mushrooms. The amanitas possess the ability to take up from the soil and concentrate the rare ductile, malleable metal vanadium, which is used to make a steel alloy high in tensile strength.

Fly Amanita (*Amanita muscaria*)

aka fly agaric, fly mushroom, bug agaric, false orange, soma, scarlet fly cup, fly brown agaric, fly poison amanita, *fliegenpilz* ("fly mushroom" in German), *mukhomor* (Russian), rod *flugsvamp* ("red fly mushroom" in Swedish)

Like all of this family, Fly Amanita starts its above-ground life in the "egg" of the universal veil. Breaking free of this skin leaves remnants resembling white warts or "butter brickle chips" on the yellow-orange cap (more red in the western U.S.). The large red-and-white speckled cap makes for a stunning picture. In fact, most cartoon mushrooms are characterizations of Fly Amanita.

You see them as lawn ornaments and salt and pepper shakers, in children's books as "toadstools," on fantasy posters—there was even one dancing to the "Nutcracker Suite" in Disney's *Fantasia*. Be careful, though, as heavy rains can wash away the warts. Caps average 6 inches across, but specimens 2 feet wide have been found. Gills are white, crowded, and free of the stem. The spore print is white. Also note the annulus ringing the stem and the below-ground volva, which, in this species, is scaly. The flesh is white, bruising to brown. In the North Woods, look for Fly Amanita under pines and aspens in June and July and then again from late August through September.

If anybody asks, tell them Fly Amanita is poisonous. Though there have been deaths recorded, it is more hallucinogenic than lethal, playing nasty tricks with mind and body. You've heard the phrase "go berserk." Did you know that this phrase was coined from the terrifying frenzied-rage of the Berserk people of Scandinavia, who were also known as the Vikings? It is believed that their fierce and inexhaustible battling was due to the stimulating influence of *Amanita muscaria*. They also raped, pillaged, and looted while under the influence. Further east in northeast Siberia, the Koryakt tribe worshiped the "mukhomor" as the "earthly incarnation of infinity, divinity, and virility." It was so highly revered that the wealthy who could afford the amanita concoction were forbidden to urinate on the ground, but, rather, were forced to go in bowls. Their urine, in turn, was drunk by the poor, who were still able to get intoxicated from it. A potent mixture may have lasted through four or five people. It is no wonder that the Fly Amanita has been associated with witches and cults through the years. (You can usually find one depicted in old paintings of witches and their brooms).

Three closely related toxins, muscimol, ibotinic acid, and muscazone, act together to produce symptoms such as apparent inebriation, confusion, incoherence, dizziness, spasms, convulsions, rage, intense desire for exaggerated physical activity (remember the Berserks?), sudden bursts of energy, deep deathlike sleep, bizarre dreams (one described mythical beasts, magic flowers, and white-robed maidens), waking to an unusual sense of well-being, and acoustical and visual hallucinations.

The last symptoms are very interesting and may have ties to Lewis Carroll's classic tale *Alice in Wonderland*. As early as 1721, there were published accounts of the cult use of *Amanita muscaria* by the Koryakt Siberians. In these reports were several references to amanita-induced macropsia, the condition where small things appear large (sound familiar, Alice?). One account of the hallucinations reads: "Some might deem a small crack to be as wide as a door and a tub of water as deep as the sea . . . All things appear to him increased in size. For instance when entering a room and stepping over the doorsill, he will raise his feet exceedingly high." Carroll most assuredly knew of these accounts. Take a look at an early version of the book. Note the mushroom that Alice is sitting on. In most editions it is a big, red-capped toadstool polka-dotted with white—none other than our hallucinogenic friend *Amanita muscaria*.

Fatal attraction

Fly Amanita's specific epithet, *muscaria*, is derived from the Latin name of the Common Housefly, *Musca domestica*, an insect pest that was poisoned using this fungus. Old-world housewives placed chunks of the sweet-smelling amanita in pans of milk, luring the flies to their death.

Polypore family—Polyporaceae

Polypore means many-pored, and it is these structures that define the family. Spores are produced and shed through tiny tubes on the undersides of the fungus, which are so tightly compacted that they appear to be a smooth surface stippled with pinpricks. A good hand lens allows one to clearly see the openings. Most species are tough and woody, perennially growing on the trunks of standing trees and downed wood. They are shelflike and stalkless. Many of you are familiar with the massive

Artist's Shelf Fungus (*Ganoderma applanatum*), whose underside provides a palette for folk artists. This species occurs in the North Woods, but mainly on deciduous trees. Remember that the polypore will always grow horizontal to the ground, so you can tell if one grew while the tree was standing or fallen. Of course, this spore-dispersing part of the "decay-machine" is only the tip of the fungus, so to speak. White, stringy fungal-roots, called mycelium, are busy breaking down the wood inside the tree. Polypores are key recyclers in the forest ecosystem.

Birch Polypore (*Piptoporus betulinus*)
aka birch fungus

This tough, spongy, stalkless bracket-style fungus is found exclusively on birches, which in the North means Paper Birch. Infested trees are usually standing, but can be dead or alive. If they're alive, though, the fungus is a sure sign that the end is near. Note the Birch Polypore's unique rolled-under margins that form a rounded rim about the pore surface. The smooth leathery skin is brownish to white and easily dented with a finger. Though an annual, it can survive hard frosts and may last well into winter. Pushing through the bark as a bulbous knob, the fungus gradually fans out. Eight-inch spans are not out of the question. The spore print is white.

Beekeepers burned this fungus to create a smoke that anesthetized the stinging critters, allowing them time to clean the hives. And if you're ever out in the woods and need a shave (legs or face) and your straightedge razor is dull, use the Birch Polypore to strop it. It really works!

Birch Conk (*Fomes fomentarius*)
aka tinder poly-pore, *amadou* (French), *frioskticka* (Swedish)

This fungus is another one of those "birch groupies," though infecting maple is not unheard of. Unlike the Birch Polypore, though, Birch Conk is a perennial, adding a new spore-tube layer each year. You can somewhat make out the annual ridges on the exterior, but if the fungus is cut in half from top to bottom, the annual spore-tubes are quite visible. Hoof-shaped, the conk grows longer each year rather than wider. Two to 8 inches is an average height range. Rings of brown, tan, and buff highlight the woody off-white exterior. Pores are minuscule, three to four per millimeter. Spore prints are white.

This is the true tinder polypore. Voyageurs and other early "pre-match" explorers would soak chunks of the tube layer in a saltpeter solution, let it dry, and then carry it along to catch sparks from their flint and steel. Ignition of fires was sure and quick.

Yellow-red Gill Polypore (*Gloeophyllum sepiarium*)
aka *vedmussling* ("wood clam" in Swedish)

This brightly colored polypore is the major agent of decay in dead conifers. It also plagues telephone poles and bridge timbers. One- to 4-inch-wide leathery fanlike brackets cluster on exposed wood. Topsides are striking, with rich brown centers edged in orange. Undersides, at first glance, appear to be gilled. But on closer inspection, note that it is just tubes and pores arranged radially in rows. This is another annual that may last into winter.

Puffball family—Lycoperdaceae

From golf ball to bowling ball in size, the puffballs are a fascinating lot. They inhabit mountaintops as comfortably as suburban lawns and change form as they ripen. Most species are quite edible when young. But all too quickly, the spores ripen and become brownish and powdery. The exterior darkens also, becoming papery. Finally a hole or tear opens in the case, and a single raindrop or tap with a curious finger sends spores flying and puts the "puff" in puffball.

**Puffball casserole
(New Jersey Mycological
Society Newsletter)**

3 cups sliced puffballs

1 cup milk

1 egg

$^1/_2$–$^3/_4$ cup
    grated Parmesan cheese

6–7 tablespoons butter

salt and pepper

$^1/_2$–$^3/_4$ cup bread crumbs

Heat oven to 350° Fahrenheit.
Layer puffball slices and bread
crumbs in a casserole dish,
dotting each layer with butter.
Beat together egg and milk and
add enough to cover. Add salt
and pepper. Spread grated
cheese over top. Bake for 20
minutes, or till knife comes
out clean.

Gem-studded Puffball (*Lycoperdon perlatum*)
aka *vårtig roksvamp* ("warty smoke mushroom" in Swedish)

This widely distributed species is probably the most common woodland puffball in North America. Golf-ball-sized and covered in conical "studs," this mushroom is distinctive. Broken-off studs leave scars on the surface. It is white on the outside and, when young, white on the inside. It appears to be stalkless, but if you brush aside some of the leaf litter you'll see a short, fat extension of the fruiting head. When the internal spore mass is white and has a cheesy consistency, the Gem-studded Puffball is edible, although one author calls them "bland at best, bitter at worst." (WARNING! The button form of the poisonous Fly Amanita can look superficially like a puffball. To test, cut it in two from top to bottom. If an amanita, it will show the cap, gills, and short stalk.) As the puffball ages, the spores ripen to yellow, then to greenish brown. The now papery skin opens at the top, allowing the escape of the spores.

Giant Puffball (*Calvatia gigantea*)

This is the really big one, the one you might confuse with a volleyball. Roundish and white with smooth skin, the Giant Puffball sits on the ground in woodlands and fields. A Minnesota specimen stood 2 feet high and weighed 45 pounds. Tales (myths?) from the 1800s describe puffballs 5 feet high and 4 feet wide. The Giant Puffball is tasty when firm and white, but when the skin dries and cracks open, it's every spore for itself. A 5-pound specimen may have upwards of seven trillion spores. A mere raindrop may "poof" a million of them skyward. It is said that if each spore matured, the combined size of the second generation would be eight hundred times the size of planet Earth!

False Morel family—Helvellaceae
Conifer False Morel (*Gyromitra esculenta*)

aka brain fungus, elephant ears, beefsteak morel, false morel, lorchel, *stenmurkla* ("stone morel" in Swedish)

Brainlike. How much more descriptive can you get? This early season (mid-May to mid-June in northern Minnesota) mushroom is brownish red and highly convoluted, just like a 2- to 4-inch-high baby brain. It is not honeycombed like a real morel (*Morchella* species), nor erect, but rather as broad as it is tall and often appearing stalkless. However, when you brush aside the pine needles and leaves, you can find the short, whitish stalk. Its flesh is very brittle, and the interior is chambered and hollow. As its name implies, look for it under conifers, especially pines, but it can also be found under aspens. Conifer False Morels are relatively long-lived for fleshy mushrooms, lasting several weeks.

The first time I encountered this species I picked it and brought it home. I was a neophyte mushroom picker and thought it might be a morel. Fortunately, I looked it up and researched its edibility. It was a lesson I'll never forget. Evidently, several deaths and many severe poisonings have been attributed to it. But its effects are quirky. In one case, two people were feasting on the fungus, but only one was poisoned. The other felt no ill effects. This narrow margin between a lethal dose and a safe one became a mycological mystery. The answer came with the discovery of the chemical nature of gyromitrin, the toxin in Conifer False Morels. It is actually monmethylhydrazine, otherwise known as MMH, a volatile and dangerous chemical that turns to a gas when heated. Shortly after a person eats a lethal dose, nausea kicks in, then vomiting, cramps, and diarrhea, followed by a high fever, jaundice, and eventual loss of muscle control. Death comes two to four days later. MMH acts directly on the central nervous system by scrambling vitamin B6, which is critical to amino acid metabolism. Tests on lab animals have shown an increase in malignant tumors.

Amazingly, MMH is used as a propellant in rocket fuels, and those accidentally exposed to it suffer symptoms identical to poisoning by *Gyromitra esculenta!* But this hydrazine toxin can be easily neutralized by cooking. With its boiling point of only 185° Fahrenheit, even sauteing will turn MMH to a gas, a form in which it decomposes rapidly. But this also means that a cook sniffing the simmering mushrooms can keel over dead while dinner guests eat heartily and live with not so much as a stomachache!

I agree with Charles McIlvaine, a mycophagist known for his willingness to eat almost any fungus, who said, "It is not probable that in our great food-giving country anyone will be narrowed to *G. esculenta* for a meal. Until such emergency arrives, the species would be better left alone."

Jelly Fungus family—Dacrymycetaceae
Orange Jelly (*Dacrymyces palmatus*)
aka fairy butter

I love the common names of fungi. Take Orange Jelly for example. It's perfect. That's exactly what it looks like: orange jelly spilled on a log or stump. And you can even eat it. Don't try spreading it on toast; rather try boiling or steaming it. But take my word, it is a fungus. The lumpy globs of orange, $^1/_2$ inch to $2^1/_2$ inches wide, are anchored to the dead wood by white stalks. You may have to lift the jelly to see this. They prefer conifer logs and, in my experience, are especially fond of cutoff stumps. The similar Witches' Butter (*Tremella mesenterica*) grows only on deciduous wood, has no white point of attachment, and tends toward yellow rather than orange. Orange Jelly in its prime is tough, but as it ages, it softens. I've seen it as late as October 23 in the canoe country, though one can find it throughout the summer.

Bolete family—Boletaceae

Two hundred species of boletes inhabit the forests of North

America. They look like your classic mushroom: a central stem growing out of the ground and topped by an umbrellalike cap. But peak under the brim of the cap and you'll discover that instead of gills they have spongy spore pores. This is characteristic of the boletes and unique in the fleshy fungi. Also, the tube layer easily separates from the cap flesh, an important trait for distinguishing boletes from the fleshy polypores. Most species are edible. The few that are poisonous have orange to red spore pores. Steer clear of any whose flesh bruises blue when handled or cut. Slugs, however, find all species to their liking. I've seen them eat the entire stalk out from under a large bolete in one night.

King Bolete (*Boletus edulis*)

aka steinpilz, cep, prawdziwek, Karljohanssvamp (Swedish)

A member of the "Royal Family of Edible Fungus," *Boletus edulis* is king of the year-round gourmet food industry. Marketed as *cep*, *steinpilz* (German), or *prawdziwek* (Polish), much of America's stock is imported from Europe in strings of dried caps. It is considered a true delicacy by Europeans. The Swedes call it *Karljohanssvamp* after King Karl XIV, who cultivated this regal fungus about the walls of his castle, Rosersberg, on the outskirts of Stockholm.

This bolete is also king-sized. A reddish brown cap, 3 to 10 inches wide, sits atop a stalk 4 to 10 inches tall. Usually the whitish spore tubes can be seen bulging below the edge of the cap. These tubes yellow with age. Spore prints are olive-brown. Brush away the needle litter to get a good look at the bulbous base of the fat-bottomed white or brownish stem. Look for the King Bolete in summer under pines, hemlocks, birch, or aspen.

The look-alike Bitter Bolete (*Tylopilus felleus*) has spore pores aging to pink and a dark pattern of netting on the stem. It also tastes very bitter.

Chanterelle family—Cantharellaceae

The typical chanterelle is yellowish orange, vase-shaped, and ground dwelling, has spore-bearing ridges instead of true gills, and is wildly delicious. As with most things in life, though, there are exceptions. While none of the Cantharellaceae are toxic, poisonings have occurred through misidentification of the Jack O' Lantern (*Omphalotus olearius*).

Golden Chanterelle (*Cantharellus cibarius*)

aka chanterelle, girolle, pfifferling, kantarellas

Golden Chanterelles start their above-ground life as anonymous round-capped mushrooms beneath the leaf litter of the forest floor. But as they push their way into daylight, the cap flattens out, eventually becoming goblet-shaped with a depression in the center. Edges of the cap are irregular and wavy. Blunt, gill-like forked ridges run from under the cap to halfway down the stem. They are highly visible. The entire fungus is yellowish (can be orange, gold, egg yellow). Caps range from $1/2$ inch to 6 inches across. A good sniff of the ridges could reveal a fruity smell that observers have described as pumpkin- or apricotlike. Look for Golden Chanterelles in July and August in the coniferous forests of the North Woods.

A fungus with a worldwide reputation must be either perniciously poisonous or perfectly palatable. In the case of Golden Chanterelles (*kantarellas* in Scandinavia and *pfifferling* in Germany), it is the latter. Their taste has been described as spicy to peppery, nothing at all like the bland mass-cultivated white things we buy in the grocery stores. They are low in protein but high in vitamin A. Pickers scour the winter California woods for the "golden goblet," which is then shipped abroad. In Sweden they are sold dehydrated in camping supply stores. Fortunately, commercial cultivation of chanterelles has eluded those who've tried. It keeps the mystery and mystique of wild mushrooms alive. As they're fibrous when raw, one should always cook them. Cream of chanterelle soup is out of this world!

Beware o' the Jack O' Lantern. It is a toxic twin to the Golden Chanterelle. Look for these three differences. Chanterelles grow out of the soil; Jacks grow on wood, though possibly buried wood. Chanterelles have blunt, forking ridges; Jacks have unforked knife-edged gills. In the North Woods, Golden Chanterelles appear in summer as individuals or in scattered groups; Jacks are mainly a September species that forms large clusters. Know thy 'shrooms or pick not the fleshy fungi!

Flask Fungi—Order Sphaeriales
Black Knot (*Apiosporina morbosa*)

The first time I saw the black lumpy growths wrapped around the bare twigs of a small tree, I couldn't figure out what had transpired. It was only later that I learned of the fungus Black Knot, which attacks several small tree species in the rose family, Chokecherry (*Prunus virginiana*) being the most susceptible. It is most noticeable in winter when the leaves are off the tree.

The visible signs of infection start with swelling bark, which soon cracks open lengthwise, revealing the diseased and yellowed wood. Slowly the wounds are encased with a hard black sheath on which the fruiting bodies grow. The young fungus knots, velvety and olive colored, eventually harden into swollen black lumpy cylinders, completely surrounding the branch for up to 12 inches. Affected twigs may be bent at odd angles and will eventually die. Small trees covered with many black knots are doomed.

Coral Fungus family—Clavariaceae

Many-branched and brightly colored, the coral fungi resemble the corals of the sea. They can be purple, pink, white, orange, and red. Most grow out of the soil, but some on logs and stumps. A few are clublike and unbranched. During the late summer and early fall in the coniferous woods of the North, this fascinating family really struts its stuff.

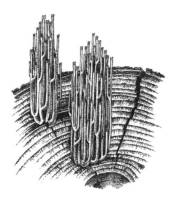

Straight-branched Coral (*Ramaria stricta*)

This honey-colored coral fungus forms tight clusters of vertical stems 2 to 4 inches tall. Each stem may branch and double-branch, but all remain perfectly parallel to one another. Varying stem height helps create a miniature pipe organ effect. This variety of Straight-branched Coral grows out of downed coniferous wood during the second half of summer and well into fall. A different variety of *R. stricta* that is grayish orange and more openly branched sprouts from deciduous logs. The spores make a golden yellow print.

Russula family—Russulaceae

"Colorful to the eye and boring to the palate" is how one jaded mycophagist described the russulas. (In fact, some are downright poisonous to the palate.) But they are nothing if not common. The two genera *Lactarius* and *Russula* are probably the most abundant and widespread of all the gilled mushrooms. Unfortunately, they are also some of the most varied and difficult to identify. They come in a bewildering panoply of reds, purples, pinks, greens, yellows, oranges, whites, and browns. One characteristic they all share is their brittle nature. If you throw a likely looking fungus against a tree and it shatters, it's a russula. The caps even snap audibly when you break them, and breaking them is not hard, for they are very fragile. Stalks fracture cleanly in two like a snapped piece of chalk. Clusters of unique large, roundish cells called sphaerocysts in among the usual long filamentous cells are responsible for the brittle nature. Caps are often wider than the fungus is tall. Spores are colorless, forming a white print.

The genus *Lactarius*, the milk caps, are so named for their production of a milky-looking latex, which beads up at the site of broken gills. Latex has been used historically in the production of rubber. Fortunately for the world, the Rubber Tree (*Hevea brasiliensis*) is a far bigger producer of latex than these fungi. Some milk caps are edible and some poisonous.

Amazingly, one toxic species (*L. torminosus*) is roasted and added to coffee in Norway and sold pickled in Scandinavia and Russia. Two hundred species inhabit the woodlands of North America.

Emetic Russula (*Russula emetica*)
aka *giftkremla* (Swedish)

This is a very common North Woods mushroom. It often grows in deep moss near rotting conifers from late summer into fall. Look for the red to reddish orange cap, striated edge, and crowded creamy white gills (occasionally forked), which are attached to the stem. At 2 to 4 inches tall, the white stalk is slightly longer than the cap is wide. Peeling the cuticle away leaves a pink stain. As the Emetic Russula ages, the cap turns up at the margin, clearly showing the parallel striations, and becomes depressed in the center. Its color fades. This species is poisonous.

Sparky says: I'm not going to give you any recipes to try because that's better left to fungus experts. But what I am going to do is put you on a path to becoming an amateur mycologist, someone who knows and respects fungi. Color, shape, size, substrate, season, and habitat can all be clues to identification, but often you need more. Spore color is one such trait that, in some situations, can safely separate similar species.

You'll need index cards (3" x 5" or 5" x 7"), a can of artist's spray fixative or clear lacquer, a fat black magic marker, a pen or pencil, and a freshly picked fungus. Blacken half an index card with the marker. Now record on the back side the species of fungus (if known), date collected, location, substrate, gill type (pores? attached? forked?), and the shape, color, size, and texture of cap, stem, and gills. Now cut the stem off your mature mushroom and lay the cap gill-side down on the index card.

Place it so half is on the black side and half is on the white side. (White or light spores will show up better on the dark side, and vice versa.) Cover with a bowl to eliminate drafts. Go to bed, and check it in the morning. Gently uncover and remove the fungus. Spray the card from 18 inches away with fixative or lacquer. Are the spores the color you thought they'd be? Check a good guidebook to key out your find. File the spore card away to help jostle your memory during future forays into the fun, fascinating, and frustrating field of fungal fingerprinting.

Family Amanitaceae (Amanitas)
- ❏ Fly Amanita   *Amanita muscaria*

Family Polyporaceae (Polypores)
- ❏ Birch Polypore   *Piptoporus betulinus*
- ❏ Birch Conk   *Fomes fomentarius*
- ❏ Yellow-red Gill Polypore   *Gloeophyllum sepiarium*

Family Lycoperdaceae (Puffballs)
- ❏ Gem-studded Puffball   *Lycoperdon perlatum*
- ❏ Giant Puffball   *Calvatia gigantea*

Family Helvellaceae (False Morels)
- ❏ Conifer False Morel   *Gyromitra esculenta*

Family Dacrymycetaceae (Jelly Fungi)
- ❏ Orange Jelly   *Dacrymyces palmatus*

Family Boletaceae (Boletes)
- ❏ King Bolete   *Boletus edulis*

Family Cantharellaceae (Chanterelles)
- ❏ Golden Chanterelle   *Cantharellus cibarius*

Order Sphaeriales (Flask Fungi)
- ❏ Black Knot   *Apiosporina morbosa*

Family Clavariaceae (Coral Fungi)
- ❏ Straight-branched Coral   *Ramaria stricta*

Family Russulaceae (Russulas)
- ❏ Emetic Russula   *Russula emetica*

Fungi of the North Woods and Boundary Waters

Appendix

North Woods Flora Habitat Finder

This habitat finder can help you, the reader, locate, in the field, the species highlighted in this book. Check off each species as you locate it. Each plant or fungus is placed under the habitat heading where it is most often found in the North Woods. In certain cases where a species is commonly found in two habitats, it is listed under both headings.

Forest

- ❑ Creeping Snowberry
- ❑ Wild Sarsaparilla
- ❑ Starflower
- ❑ Red Baneberry
- ❑ False Lily-of-the-Valley
- ❑ Pink Ladyslipper
- ❑ Wild Columbine
- ❑ Bunchberry
- ❑ Clintonia
- ❑ Twinflower
- ❑ Rose Twisted Stalk
- ❑ All pyrolas
- ❑ Spotted Coralroot
- ❑ One-flowered Wintergreen
- ❑ Pipsissewa
- ❑ Indian-Pipe
- ❑ Large-leaved Aster
- ❑ Wintergreen
- ❑ Beaked Hazel
- ❑ Red Osier Dogwood
- ❑ Paper Birch
- ❑ Quaking Aspen
- ❑ Northern White Cedar
- ❑ Balsam Fir
- ❑ Jack Pine
- ❑ Red Pine
- ❑ White Pine
- ❑ Sensitive Fern
- ❑ Cinnamon Fern
- ❑ Interrupted Fern
- ❑ Ostrich Fern
- ❑ Wood-Horsetail
- ❑ Ground-Cedar
- ❑ Round-branched Ground-Pine
- ❑ Stiff Clubmoss
- ❑ Shining Clubmoss
- ❑ Running Clubmoss
- ❑ Conifer False Morel
- ❑ Gem-studded Puffball
- ❑ King Bolete
- ❑ Fly Amanita
- ❑ Golden Chanterelle
- ❑ Straight-branched Coral
- ❑ Emetic Russula

Sunny openings/old burns

- ❑ Wild Strawberry
- ❑ Lowbush Blueberry
- ❑ Prickly Wild Rose
- ❑ Wild Raspberry
- ❑ Thimbleberry
- ❑ Common Evening-Primrose
- ❑ Fireweed
- ❑ Spotted Joe-Pye Weed
- ❑ Juneberries
- ❑ Mountain Maple
- ❑ Quaking Aspen (young)
- ❑ Jack Pine (young)
- ❑ Bracken Fern

❏ Field-Horsetail
❏ British Soldier
❏ Pixie Cups
❏ Reindeer lichen
❏ Giant Puffball

On trees

❏ Eastern Dwarf Mistletoe
❏ Black Knot
❏ Birch Conk
❏ Birch Fungus
❏ Yellow-red Gill Polypore
❏ Orange Jelly
❏ Old Man's Beards

Rock faces and cliffs

❏ Bearberry
❏ Harebell
❏ Pin Cherry
❏ American Mountain-Ash
❏ Rock Polypody
❏ British Soldiers
❏ Pixie Cups
❏ Reindeer Lichens
❏ Rock Tripes

Shoreline

❏ Sweet Gale
❏ Wild Iris
❏ Northern White Cedar
❏ Royal Fern

Wet areas

❏ Marsh Marigold
❏ Spotted Joe-Pye Weed
❏ Speckled Alder
❏ Black Ash

Bogs

❏ Leatherleaf
❏ Labrador-Tea
❏ Pitcher-Plant
❏ Small-fruited Bog Cranberry
❏ Round-leaved Sundew
❏ Tamarack
❏ Black Spruce
❏ Sphagnum mosses

Lakes

❏ Marsh Marigold
❏ Bullhead Water-Lily
❏ Fragrant White Water-Lily
❏ Wild Iris
❏ Broad-leaved Arrowhead
❏ Water-Horsetail

Wildflower Bloom Phenology

page		May 1 2 3 4	June 1 2 3 4	July 1 2 3 4	Aug. 1 2 3 4	Sept. 1 2 3 4	Oct. 1 2 3 4
48	Sweet Gale						
107	Wild Strawberry						
57	Leatherleaf						
59	Bearberry						
80	Marsh Marigold						
61	Lowbush Blueberry						
94	Creeping Snowberry						
123	Wild Sarsaparilla						
106	Starflower						
82	Red Baneberry						
140	False Lily-of-the-Valley						
144	Pink Ladyslipper						
84	Wild Columbine						
118	Bunchberry						
136	Clintonia						
54	Labrador-Tea						
64	Prickly Wild Rose						
127	Twinflower						
109	Wild Raspberry						
138	Rose Twisted Stalk						
74	Bullhead Water-Lily						
100	The pyrolas						
77	Fragrant White Water-Lily						
147	Spotted Coralroot						
142	Wild Iris						
86	Pitcher-Plant						
111	Thimbleberry						
95	Small-fruited Bog Cranberry						
102	One-flowered Wintergreen						
115	Common Evening-Primrose						
100	One-sided Pyrola						
125	Harebell						
113	Fireweed						
98	Pipsissewa						
134	Broad-leaved Arrowhead						
104	Indian-Pipe						
132	Spotted Joe-Pye Weed						
89	Round-leaved Sundew						
130	Large-leaved Aster						
92	Wintergreen						

Glossary

achene - dry, hard nonsplitting fruit with one seed loosely attached inside

actinomycetes - the bacteria responsible for nitrogen fixation in plants other than legumes

anaerobic - referring to cells that can live without oxygen

angiosperm - the group of plants whose seeds are borne in a mature ovary (fruit)

annulus - in gill fungi, the ring around the stalk which is the remnant of the inner veil

anther - The stamen's enlarged terminal end which holds the pollen

anthocyanin - blue, purple, or red pigments found in cell sap

aril - an extra seed covering, often brightly colored, which may attract animals who then disperse the seed

basal rosette - a ring of leaves found at the base of a plant stem

bract - modified leaves of green or other colors, associated with the flower

catkins - a hanging spike of tiny unisexual flowers on some woody plants

clone - a population of individual plants descended by mitotic division from a single ancestor

corolla - the name used to describe the petals of a flower as a unit

cotyledon - a seed leaf, which stores food in dicots and absorbs food in monocots

dicots - the group of plants characterized by an embryo with two cotyledons or divisions, netlike leaf veins, and flower parts in fours or fives

disk flower - the tiny insignificant flowers that en masse form the central flower head in composites such as sunflowers and daisies

drupes - a fleshy fruit such as a cherry, olive, plum with a hard nut or stone in the center

elaters - winglike structures on horsetail spores that presumably aid in dispersal

endosperm - tissue containing stored food that surrounds the developing embryo and fuels seed growth

epidermal cells - cells found in the outermost layer of leaves, young stems, and roots

epiparasite - a parasite that grows on its host

family - the taxonomic group between order and genus

floret - One of the tiny individual flowers that make up a composite flower, such as a daisy

fronds - the leaf of a fern

gametophyte - in ferns, the structure one cell layer thick, from which sperm and egg are released and from which the fern's growth originates

genus - the taxonomic group between family and species; the first name in a plant's Latin or scientific name

hibernacula - winter buds

hyphae - a single tubular filament that acts as fungal root directly absorb nutrients (see mycelium).

inflorescence - the entire flower cluster

legume - a member of the pen or bean family (Fabaceae)

lenticels - spongy openings in the stem or trunk of plants that allow the exchange of gases

monocots - the group of plants characterized by an embryo with one cotyledon, parallel leaf venation, and flower parts in threes

mycelium - a mass of hyphae that together form the body of a fungus

ovary - the swollen base of the pistil, in which seeds develop

palisade tissue - tissue composed of upright columnlike cells

petioles - the stalk of the leaf

phloem - the food conducting tissue in plants

pinnae - primary leaflet of a frond

pinnules - pinnae divided into even smaller leaflets

pistil - central female organ of the flower

pollinium - an orchid's sticky pollen-bearing package

pome - a simple fleshy fruit, in which the outer portion is formed by the floral parts that surround the ovary and expand with the growing fruit such as an apple or pear

raceme - a longish cluster of flowers arranged singly along a stalk, each flower with its own small stalk

ray flowers - flowers with showy large petals borne on the circumference of composite flowers such as daisies

receptacle - the part of the flower stalk that bears the floral organs

rhizone - a horizontal underground stem

samara - a winged fruit with one seed such as ash or elm fruit

saprophyte - organism that depends on decaying organic matter for nutrition rather than photosynthesis

senescence - the plant stage from full maturity to death

sepal - a small modified leaf usually growing below the actual petals of the flower

sessile -without a stalk

soredia - specialized reproductive units of lichens made up of a few algal cells surrounded by fungal hyphae

sori - clusters of sporangia

species - a specific kind of organism that can be identified by a binomial scientific name

sporangia - hollow structures in which spores are produced

sporophyte - the spore-producing phase in a life cycle of a plant

stamen - the male flower organ (usually several per flower) which is tipped by the pollen-bearing anther

stigma - the tip of the pistil

stolon - a stem such as a strawberry runner that grows horizontally on the ground surface

stomata - minute openings that are used for gas exchange in the epidermal cell layer of leaves and stems

strobile - a reproductive structure composed of a group of modified leaves at the end of a stem

style - the stalk of the pistil arising from the ovary

symbiosis - two dissimilar plants living together in close association

tripinnate-pinnatifid - fern leaves or fronds that are divided into leaflets, leaflets divided twice again

vestigial - once functional, now useless, organism part

volva - cuplike structure at the base of the stem of certain fungi

whorl - three or more leaves radiating from a point on the stem

xylem - plant tissue that conducts water and minerals

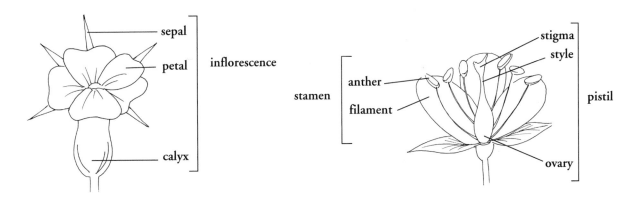

sepal

petal

inflorescence

calyx

stamen

anther

filament

stigma

style

pistil

ovary

Flower parts

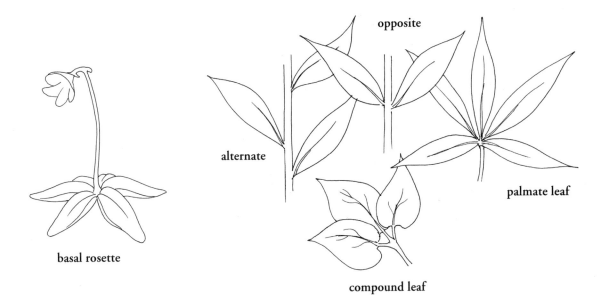

opposite

alternate

palmate leaf

basal rosette

compound leaf

Leaf arrangements

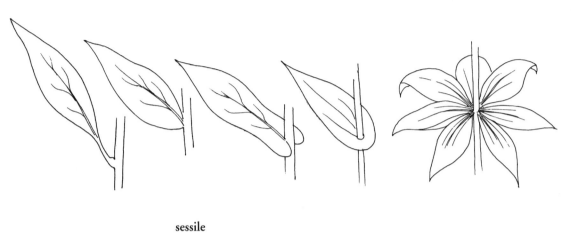

petiole   sessile
(stalkless)   clasping   perfoliate   whorl

Leaf attachment

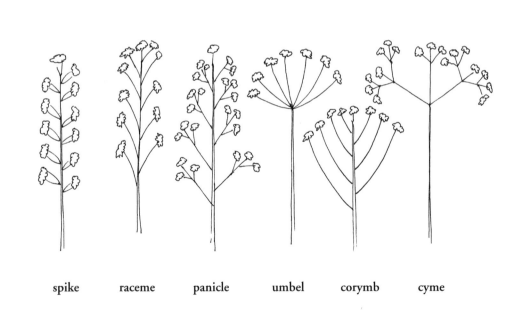

spike   raceme   panicle   umbel   corymb   cyme

Flower arrangements

Mark "Sparky" Stensaas is a freelance naturalist, photographer, and writer. He is the author of *Rock Picker's Guide to Lake Superior's North Shore, Wildflowers of the BWCA and the North Shore,* and *Canoe Country Wildlife,* also published by the University of Minnesota Press. He lives in Minnesota's Carlton County.

/